Integrating Design and Test:
Using CAE Tools for ATE Programming

KENNETH P. PARKER
Hewlett-Packard Company

COMPUTER SOCIETY NUMBER 788
LIBRARY OF CONGRESS NUMBER 87-70816
IEEE CATALOG NUMBER EH0261-8
ISBN 0-8186-8788-6
SAN 264-620X

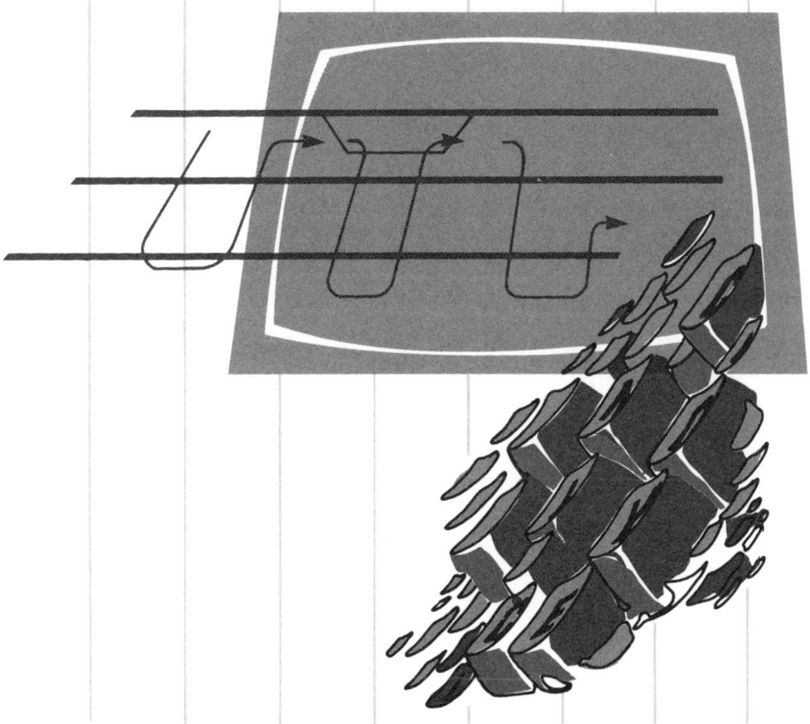

THE COMPUTER SOCIETY OF THE IEEE

IEEE
THE INSTITUTE OF ELECTRICAL AND ELECTRONICS ENGINEERS, INC.

IEEE COMPUTER SOCIETY PRESS

INTEGRATING DESIGN AND TEST:

USING CAE TOOLS FOR ATE PROGRAMMING

KENNETH P. PARKER

Manufacturing Test Division
Hewlett-Packard Company

Published by Computer Society Press of the IEEE
1730 Massachusetts Avenue, N. W.
Washington, D. C. 20036-1903

Cover designed by Jack I. Ballestero

Sponsored by Test Techology Technical Committee

Printed in the United States of America.

Computer Society of the IEEE Order Number 788*
Library of Congress Number 87-70816
IEEE Catalog Number EH0261-8[1]
ISBN 0-8186-8788-6
SAN 264-620X

Order from:

Computer Society of IEEE*
Terminal Annex
P.O. Box 04699
Los Angeles, CA 90051

IEEE Service Center[1]
445 Hoes Lane
P.O. Box 1331
Piscataway, NJ 08855-1331

Computer Society of the IEEE*
13 Avenue de l'Aquilon
B-1200 Brussels
BELGIUM

IEEE

THE INSTITUTE OF ELECTRICAL AND ELECTRONICS ENGINEERS, INC.

Preface

In the early months of 1986, I was given the task of 'linking' the output of a digital simulator from an outside vendor to an ATE system produced here. This assignment grew into a more general effort to address the many CAE systems that are now available. My first attempt to define a simple transformation of syntax between simulator and ATE quickly proved that there was little probability of success. The reason was that digital simulations are rarely targeted at the requirements of testing nor do they observe the practical constraints imposed on ATE systems by laws of physics and economics. This is ironic because CAE systems have tremendous potential for synergism with ATE systems. It became clear that a text was needed that specifically addressed this issue, in a form suited to practitioners using either CAE or ATE systems. This work attempts to fill this need.

I would like to note the following people who have helped produce this book. Drafts were reviewed up to three times by John Algiere, Ray Balzer, Wayne Chism, Irv Christy, Mike Dapron, Chuck Ely, John Ferguson, Mike Gravitz, John Keller, Phil King, Doug Manley, Kim Mast, Rick Palmer, Rick Robinson, Brian Smith, Ted Turner, and Gary Zeigler. Any errors that survived their careful efforts are my own.

I would also like to thank those who have provided support, assistance, editing, artwork, photographs, aid, and encouragement. They are Dick Barney, Jan Berry, Margaret Brown, Jon Butler, Bill Carroll, Gary Guerriero, John Harding, Chuck House, Gary McCormack, Bloor Redding, Shirley Shirk, John Scruggs, Chip Stockton, and Jack Trautman.

Finally, I want to thank my wife Jana for her patience during the weekends and nights that were spent on this project.

Ken Parker

Loveland, Colorado
April 1987

CONTENTS

INTEGRATING DESIGN AND TEST: USING CAE TOOLS FOR ATE PROGRAMMING

Chapter 1
Introduction

Chapter 1

Introduction

This book addresses several different readers. Foremost is the digital circuit designer who is using computer aided engineering (CAE) tools. Increasingly, the designer is called upon to enhance the testability of a design and even to produce tests. These new responsibilities are outside the training normally provided to designers both in schools and on the job. This book is intended to fill this gap. With it, the designer can develop a model of a test system, ask the right questions about the parameters of its operation, develop a test strategy, and better understand those aspects of a design that affect testability.

Test engineers utilizing automatic test equipment (ATE) should also find this book helpful because it provides a vehicle for them to influence designers. Many companies still have design and test departments that are organizationally isolated and are even separated by hundreds of miles. Traditionally, testing has been a 'downstream' activity providing little or no feedback to the design process. As design and test become more closely related, test engineers should begin to see the same CAE tools become available to them. Ultimately, they will be networked together with the designers, sharing a common database.

A third constituency consists of managers who are grappling with the concept of 'linking' design and test. Linking involves far more than running a cable between the two departments. Responsibilities are shifting, new efficiencies are being sought, and new performance measures must be put in place. With this book, design and production managers should be able to develop a common vocabulary and a strategy for effectively utilizing the wonderful new design and production tools that are available.

Yet another audience comprises vendors of CAE systems. They too have only recently discovered the synergism that their systems have with test equipment. At this writing, we are seeing the first generation of 'links,' made up primarily of data translation and physical transmission protocols, becoming available. Future generations will incorporate much more knowledge of the general testing problem and the specifics of test equipment. CAE systems will eventually become design and test 'experts.'

Finally, it is hoped that students of computer engineering curricula will be given exposure to the practical problems addressed here. While this book is not intended as a rigorously theoretical work, it deals with basic engineering issues and often must relate these to fundamental principles. The simple examples given here can be developed into laboratory exercises to illustrate these issues.

1.1. A Brief History of the Design Process

In all but the smallest companies, there is a division of labor between design and production, where testing is done. A high degree of specialization has resulted, with each department largely unaware of the problems and concerns of the other. The design department has been thought of as 'upstream' from the production department. When the design department was finished with its part of the development process, the result was passed on to production. A euphemism for this was to 'throw the design over the wall' into production. While not a pleasant image, it accurately reflects the attitudes of many of the people involved and their managers; design and production are separate entities that cooperate only minimally.

1.1.1. Manual Design

Only a few years ago, circuit designs were done manually, with pencil, paper, blueprints, etc. An important part of engineering curricula was draftmanship. The circuits designed were quite simple and often analog. They were physically breadboarded and analyzed with benchtop instrumentation. A few simple experiments were conducted to improve the design. Often, the breadboard and some blueprints were given to the production department where additional work was done to make the design practical to build in quantity. This was fraught with risk since much of the original designer's knowledge was lost in the transfer. Mistakes and delays were common.

The digital revolution, which started with small scale integration and sky rocketed to ever higher complexities, placed a huge amount of strain on this process. Perhaps you remember the first transistor radios that competed based on the number of transistors they contained (usually fewer than 10). Now, digital radios are produced containing countless thousands of transistors in packages of similar size. Designs like these coming 'over the wall' are certainly difficult to move into production. New tools and more cooperation between departments are necessary.

1.1.2. The Emergence of the Logic Simulator

Very early, it was recognized that exhaustive analysis of digital logic was impractical. In the mid 1960s, digital logic simulators began to appear. (See Chapter 4 for a description of several kinds of simulators that have evolved.) It is interesting to note that the first were dedicated to developing tests. They were cumbersome, slow, and expensive to run, but highly valuable.

What value do simulators provide? First of all, they provide a means to analyze a complex digital design. They manage the huge volume of detail and data needed for this analysis. They protect against human error, a very real threat in complex designs. But more importantly, they allow a designer to try a larger number of experiments, modifications, and performance verifications than would be possible with a physical prototype, in a far shorter time. Every time a design error is uncovered with a simulation, it represents the potential saving of a design turnaround of a physical prototype. Also note that one can overlap the development of parts of a circuit. For example, an integrated circuit (IC) can be developed and verified with a simulator by one designer while another uses the IC model to simulate its behavior in a circuit board. Changes can be quickly tried and verified both at IC and at board levels. All of this occurs without the existence of actual parts. This shortens the development cycle, which is today's competitive imperative.

1.1.3. The Emergence of CAE

Digital logic simulators and a collection of other tools have matured to the point where they are packaged and offered as computer aided engineering tools. Combined with inexpensive computing platforms, we now have CAE workstations available, as discussed in Chapter 3. It is important to note (and it will be stressed several times) that CAE was born in the design part of the development process. The prime users today of CAE systems are designers. ATE systems, however, came into being along a different path; that is, in the production test area. The prime users of ATE today are production test engineers. The first use today of CAE systems is for *design verification,* not for *test development.* It is crucial to recognize the difference, and that is the reason this book exists. CAE tools can be used for test development, *if properly applied.*

1.2. A Brief History of ATE

Automatic test equipment has been around for over 20 years, first as internally developed testers and now as commercially available equipment from a variety of vendors. The most forceful driver of the development of ATE has been the rise of integrated circuitry. Higher levels of integration have made IC test a very difficult job. For ICs to be cost effective, testing must be fast and thorough. Test times of only a few seconds per IC are common. It is often speculated that ultimately testing will be the limiting factor in the evolution of IC technology. For the same reasons, testing at the board level is becoming increasingly difficult.

1.2.1. The Emergence of Automated Functional Test

The first digital testers performed 'functional tests' (i.e., they stimulated and observed circuit responses through the circuit's I/O ports). IC test is essentially always done this way, while, as we shall see, board test has added some alternate approaches.

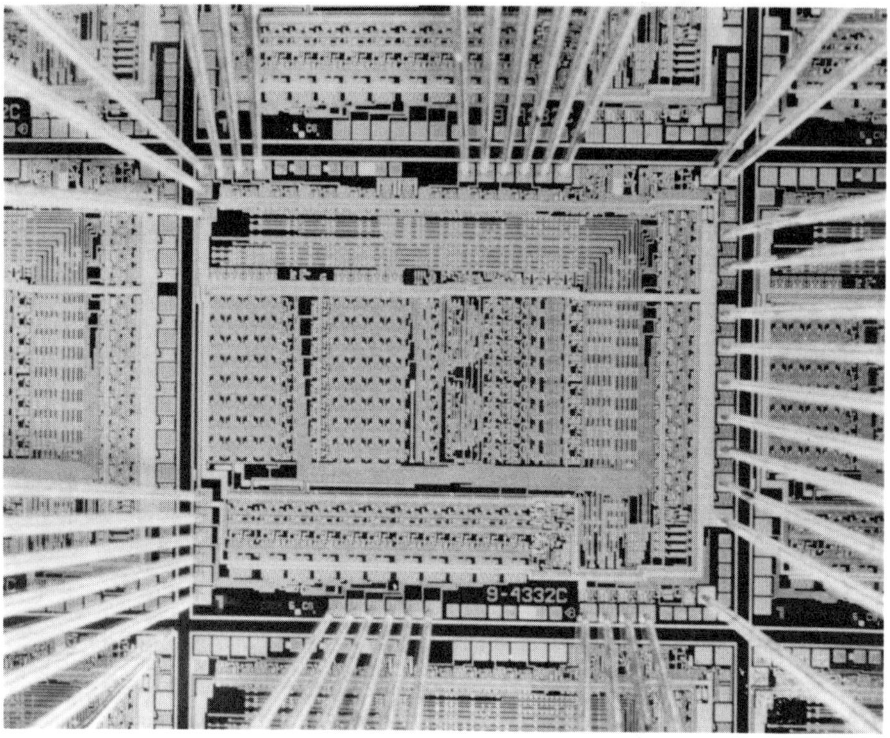

Photomicrograph of an integrated circuit under test. The IC is one of many fabricated together on a wafer of silicon. After this test, called wafer sort, the ICs are cut apart, and only those that pass the test are placed in IC packages. After packaging, they will be tested again. The probes used to contact the IC are significantly narrower than sewing needles. The IC shown is approximately 4x5 mm and is representative of mature (old) technology.

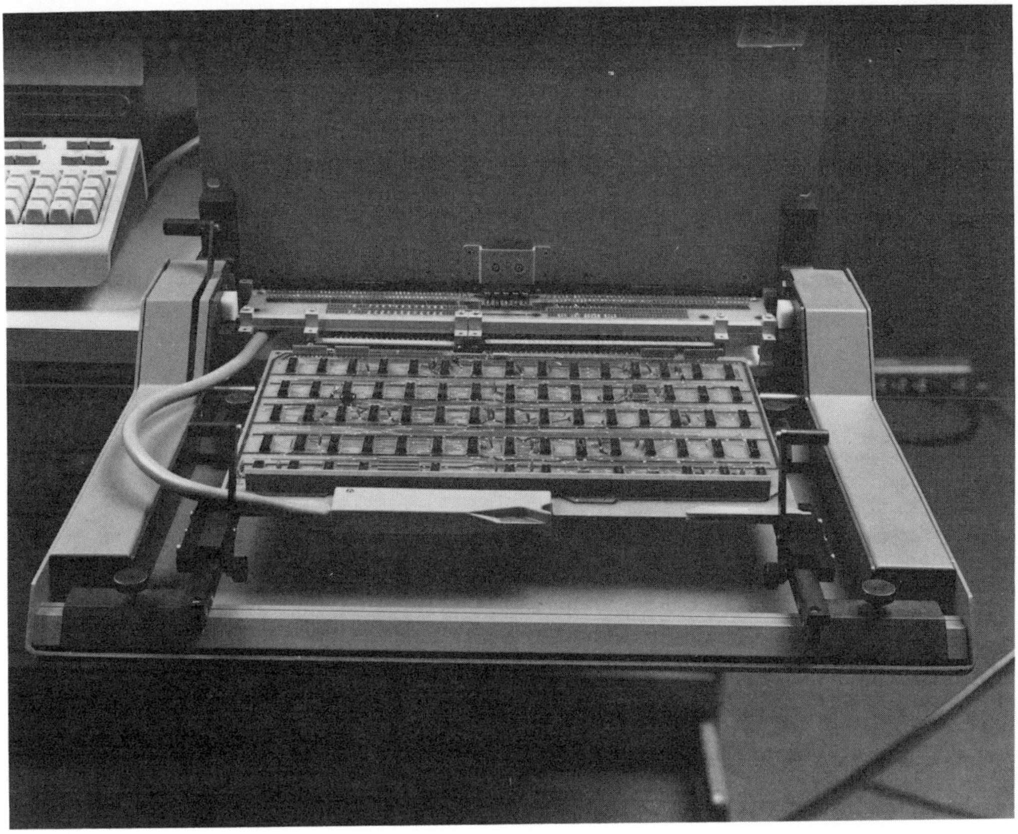

Photograph of a printed circuit board undergoing edge connector functional test. The board is plugged into an adaptor that translates the tester's physical connector configuration into that of the board's. Since this board has a second edge connector in front, a cable with hood is connected to provide tester access to it.

Meaningful analog tests are difficult to accomplish this way, requiring highly customized equipment essentially unique to each test situation. Therefore, ATE for analog functional test is largely custom built.

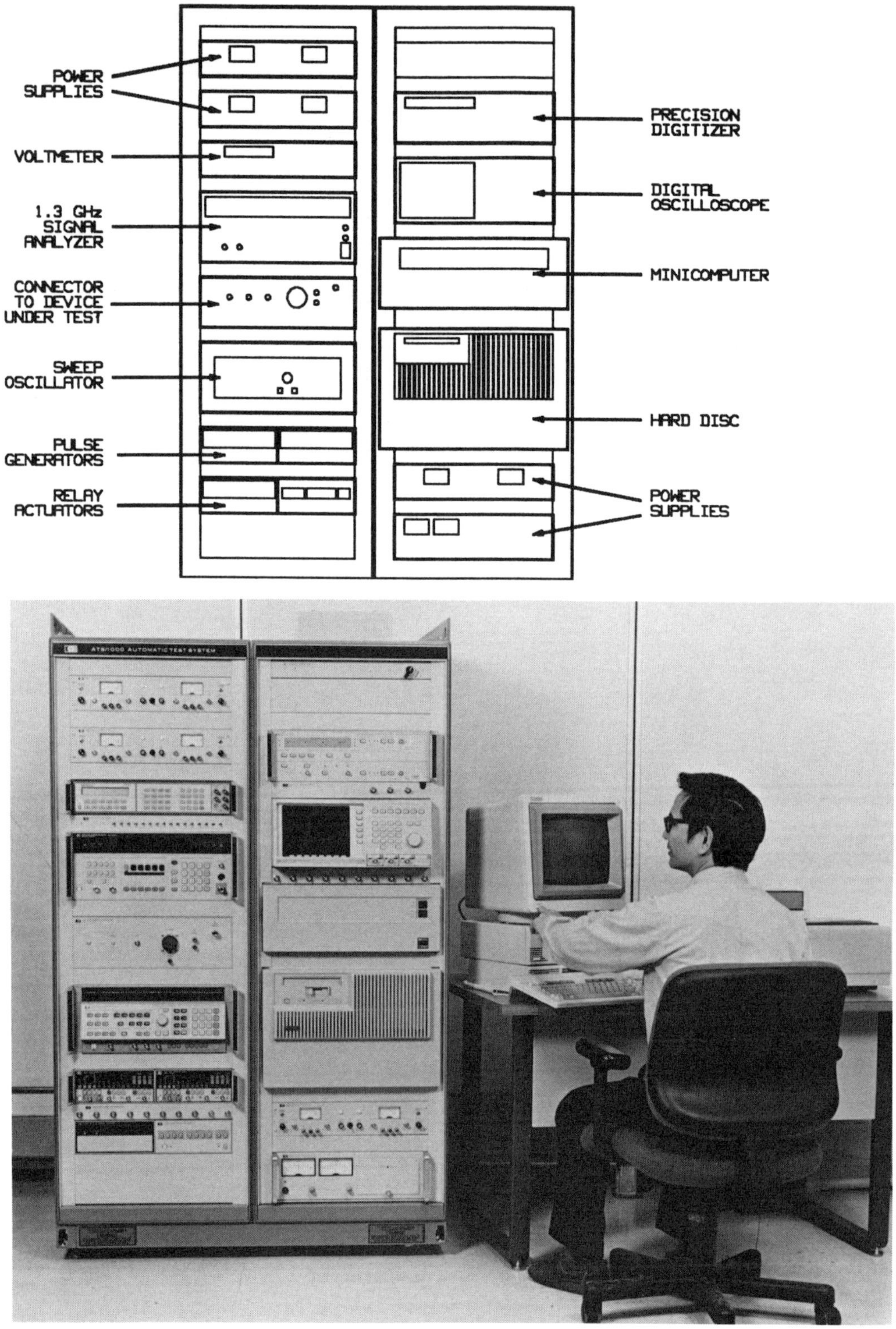

Photograph and locator diagram of a custom built analog test system for logarithmic amplifiers. Note the variety of analog measurement instruments, all under control of the minicomputer.

Commercial digital functional testers, however, quickly caught on and some of the first utilized digital logic simulators as test preparation tools were developed *more than 15 years ago*. This significantly predates the advent of CAE workstations. As circuit integration progressed from small scale through large scale and beyond, it became increasingly difficult to simulate circuit boards on existing simulators. One of the toughest problems cropped up as medium–scale integration gave way to large–scale integration: the lack of adequate circuit models. Without good models, simulation became nearly impossible. For example, the 74181 four–bit ALU documentation provides a complete gate level description. This was not to be the case with the first microprocessor or its descendants. LSI parts are described on a higher, functional plane. Indeed, this author was quoted in 1979 as saying, 'You can now buy more gates with less pages of specifications than at any time in history.'[1] As accurate gate level models became scarce, new approaches were explored for testing boards.

1.2.2. In–Circuit Test

Roughly 10 years ago, a new testing technique appeared that eliminated the need for logic simulation as a tool for preparing digital tests. Indeed, program preparation became so simplified, it could be automated to a high degree. Even better, it allowed accurate and effective tests of analog and hybrid circuits as well. These had always been a problem for functional testers. The technique is known as in–circuit (or in situ) testing.

The key to in–circuit testing is *access*. The bed–of–nails fixture allows electrical access to each node (also called nets or traces) of the circuit. Then, by a process of isolation by guarding, individual components can be tested as if they were standing alone. Both analog and digital components are addressed this way, although the isolation technique differs. For a complete technical description of these techniques, see references [Crook 1979, Balliew 1984, GenRad 1984, Harwood 1984].

In–circuit testing effectively revolutionized the art of testing. The cost and time for preparing a test was dramatically reduced. The high nodal access allowed excellent diagnostic resolution. The testing problem appeared to be tamed until new technological forces entered the scene.

[1] *Electronics Test*, March 1979, p.65. (The author regrets the lack of grammatical polish in this quote.)

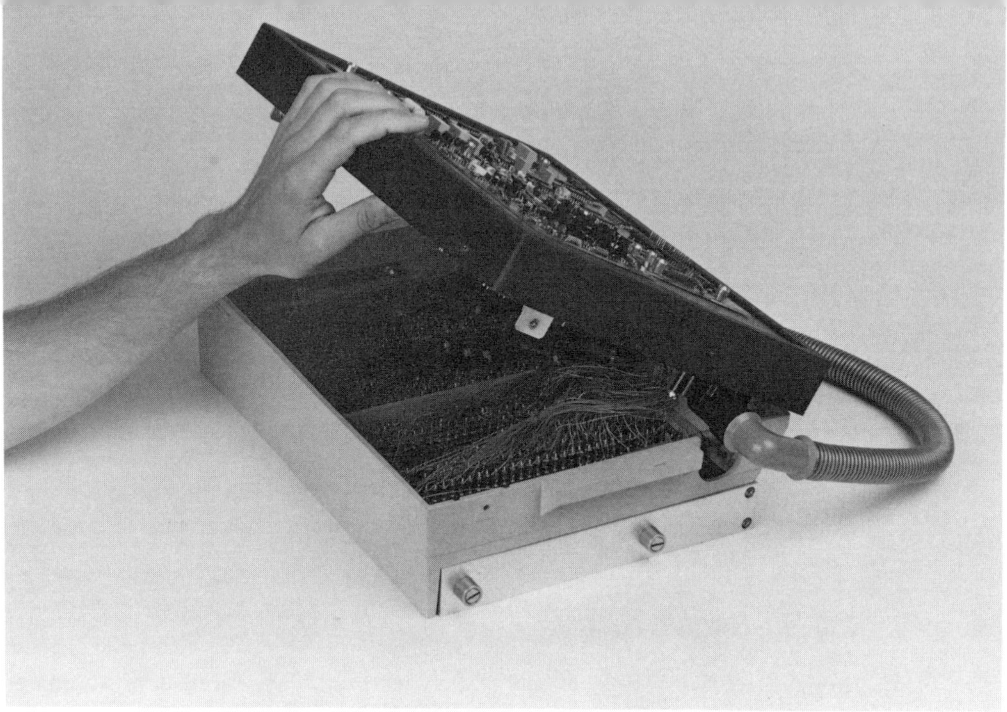

Photograph of an in–circuit test fixture. The fixture is opened to show the interconnect between the patch–panel plane and the upper platen structure where the spring loaded nails are located. The array of patch–panel plugs forms the interface to the tester electronics. The board under test is shown in place on top of the fixture. The hose supplies approximately 0.1 atmosphere of vacuum that both holds the board in place (guided by tooling pins) and compresses the movable platens that hold the nails. The nails make contact with nodes on the underside of the board, providing in–circuit access.

Since electrical access is so necessary for the success of in–circuit testing, technological changes that diminish access are a threat. This is precisely what is happening today with the revolution in device packaging and interconnect technology. 'Surface mount' technology allows the mounting of components on boards (even on both sides of boards) without the need for 'through–holes.' This means that nodal access to such components may require nail access to both sides of a board, which is a difficult mechanical proposition. Another factor is the development of denser packages with pin spacings that are significantly smaller. Again, placement of nails in dense clusters is a mechanical challenge. Advanced hybridization techniques are placing many analog and digital components on a ceramic substrate that is then encapsulated, removing accessibility.

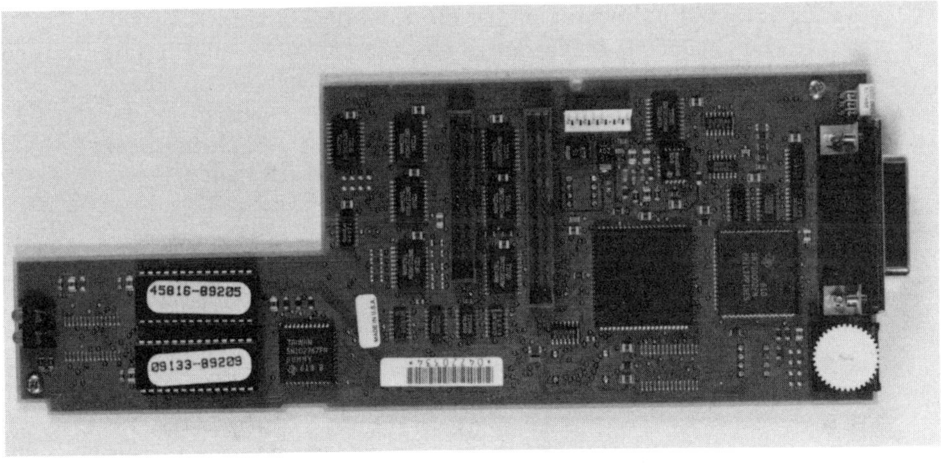

Photographs of boards using older 'through-hole' technology (top) and newer 'surface-mount' techniques (bottom). Using pure surface-mount technology, holes will only exist to provide pathways (vias) between layers of circuitry. Some signals may not be visible from the bottom side. Through-hole boards have a hole for every component lead (as well as vias), thus assuring that all signals are accessible from the bottom side.

Other problems are related to operating frequencies, which are growing steadily higher. Bed–of–nail fixtures suffer from bandwidth limitations. Test lengths are growing, reducing throughput, and stressing the memory limits of testers.

1.2.3. Combinational Testing

In recognition of the limitations of in–circuit testing, there is renewed interest in functional test techniques for board test. A new and very recent trend is to integrate both in–circuit and functional test capabilities into a single tester, called the combinational tester. This addresses access limitations by providing an alternate means of testing and allows a user to do a final functional test (sometimes called an at–speed test). This is to see if there are any unwanted interactions among devices on a board that, by definition, would not be visible during in–circuit testing. These interactions are difficult to characterize and are often indicative of marginal circuit design. Nevertheless, the desire exists to perform functional tests even after thorough in–circuit testing.

1.3. The Future of Design and Test

What does the future hold for design and test? One thing seems clear; design and test, though residing in different departments, must become more closely coordinated. The first evidence of this is the volume of literature on design for testability (DFT). (See the bibliography in [Williams 1983].) The idea that a designer has a great deal of leverage on the test problem is now widely accepted. However, much more must be accomplished. Much DFT literature talks mostly about design and not so much about testing. In practical situations, a designer must be familiar with the test equipment to be used in production so that informed tradeoffs can be made. For example, if in–circuit test is the primary method that the production department will use to test a circuit board, then a designer should be very conscious of nodal access issues, among other things. If these prove to be too restrictive, then appropriate changes must be made either in the design goals or in the production strategy. Such changes will likely not come easily or quickly and will definitely require attention from higher levels of management.

1.3.1. Capturing, Preserving, and Sharing Data

Perhaps the foremost contribution of CAE to the design and test processes is its ability to capture huge amounts of highly relevant data. Furthermore, data are preserved, easily accessible, and shareable. This contribution is only now being exploited. Right now, the data available from CAE workstations are aimed at circuit designers; that is, they are indexed, organized, and presented in ways that appeal to designers. As time goes on, data relevant to other constituencies, notably test engineers, will become more easily and directly available. It is quite reasonable to expect that, in the future, design and production departments will share the entire design database over an appropriate network. This raises a strategic management issue having to do with how networks deal with distance and data rates. It may be that higher data rates such as are available with local area networks will be necessary. If so, then design and production departments will have to be much closer than the intercontinental distances sometimes observed today.

1.3.2. Design Verification versus Test

In the future, a stronger distinction needs to be made between design verification and testing. The production department is charged with volume manufacture of high quality and reliable products. Designers are charged with creating the designs of these products and verifying that all important design criteria have been met. The two activities are quite different, and yet one often finds that test engineers are trying to make design verification tests (thrown over the wall) work on their test equipment. This effort is nearly always doomed to failure. (Remedies are discussed in Chapter 5.) Designers and their management must become aware of the requirements of testing so that they can assure that bona–fide production tests are delivered to the test department. Chapter 2 provides a start in this direction. Well–meaning designers who portray their verification test files as suitable for production testing may actually be delaying the development cycle, as time and effort are wasted trying to make the tests work, and test yields suffer.

1.3.3. Streamlining the Development Cycle

A subject often receiving attention today is how to reduce design cycle time. This misses the point. It is true that the new CAE tools may help make designing easier and reduce unnecessary turnarounds. However, if CAE tools are properly applied to design *and* test, then they can reduce the overall *development* time.

It is becoming a common complaint that developing tests for circuits can take longer than the original design. Then, to overlook test development time when applying CAE tools is to overlook about 50 percent of the opportunity for improvement. As was stated before, synergism exists between the two efforts that can benefit from viewing the two together. While each can be improved by itself, there can be better results by coordinating the efforts.

1.4. Management Goals

If you are a manager engaged in 'linking' design with test, what are your goals? A few possibilities are outlined below. Perhaps some of them are inconsistent with special situations in which you find yourself. For example, the sharing of design data brings a new set of problems to those worried about data security. If so, then the disadvantages of not sharing data must be weighed against security issues. Perhaps capital resources are plentiful, but qualified people are unavailable. Are your volumes low but yield requirements extremely high? Vice–versa? Here are some goals that should cover a spectrum of situations.

1.4.1. Better Quality, Lower Costs, Faster Development

These goals are obvious, or are they? These are interrelated and possibly conflicting goals. How to choose among them? How to speed product development without throwing money at it and without sacrificing quality?

A popular sentiment today is that the 'time–to–market' (i.e., product development time) must get shorter or you won't survive. CAE tools and ATE seem like naturals for this. However, experience has shown that misapplication of these tools can cause delays. Pundits have named this result 'Computer Aided Delay.'

Customers are expecting higher levels of product quality. It seems that CAE, with all of its thoroughness, should allow an increase in quality. However, drops in product yield could be the reward. How? By confusing design verification with test development. Are tests being created for the proper fault spectrum?

Buying a roomful of CAE workstations will not help if the designers and test engineers do not know how to use them toward the larger goal of integrating design and test. The biggest jobs will be education and coordination. Create a flow chart of the *entire* development process, from product inception to mature production, even to field support or repair and obsolescence. Ask where CAE and ATE need to cooperate to capture, preserve, and share data. Find where turnarounds and delays occur. Figure out how to maximize the probability that a subtask will succeed on the first try. How do the new tools fit in?

1.4.2. Rationalized Design for Testability

Design for testability is often espoused but not often managed rationally [Parker 1986]. Testable design techniques are often specified without taking into consideration how a circuit will actually be tested. What should be done first is to analyze the test capabilities expected to be used in production. This should be taken into account even before major decisions such as packaging, circuit partitioning, and technology choices are made. When these decisions are made, it may be necessary to update or change the way testing is done or to change the product's parameters appropriately.

IC designers have, to a large degree, already taken this advice to heart. This is because they have been bloodied in the past by technology advances that made design for testability mandatory, and there is precious little choice in how testing is accomplished. It is true, however, that the techniques they have applied to make IC testing easier will not enhance board test as much, *if at all* [Parker 1986]. In a vertically organized company or one using custom ICs (or application specific ICs, ASICs), an opportunity exists to ask the question, 'How can I design this IC to make an impact on testability at board test?' Of course, the type of board testing projected has a large effect on the answer.

1.4.3. Creation of a Living Development Process

In the past, the development cycle was looked upon as a 'feedforward' process. A product was conceived, then designed, then fabricated, then tested (and fixed), and, finally, shipped to customers. There was little 'feedback.' The quality of the product was adjusted to meet minimum requirements by weeding out bad units and sending them to a repair area. Seldom was it asked, 'Why did a unit fail and how could we prevent a reoccurrence?' The entire process was static, unchanging, inflexible, by–the–book, etc.

Today, we are moving toward a new philosophy of producing products. We hear new terms such as 'total quality control' (TQC) [Feigenbaum 1983], 'statistical quality control' (SQC) [West.Elec. 1956], 'just in time' (JIT) manufacturing [Hall 1985], and the like. The Japanese have implemented many of these concepts on a national scale. Ironically, some of these are originally American ideas that are only now coming into practice in North America [Peters 1984] . These concepts are a means to achieve an end, here termed a 'living development process.'

How does a development process 'live?' Living processes are dynamic, adaptive, and self optimizing. The 'dynamic' attribute means they have the ability to tolerate, even thrive on, change. 'Adaptive' means that they can react to conditions imposed by the surrounding environment. 'Self optimizing' means that they adapt in such a way as to improve performance. To possess these attibutes, a process needs mechanisms for observing, measuring, quantifying, and judging the performance of every step of the process. Information must be gathered, collated, analyzed, and then fed, backward or forward, to each place in the process where changes are effected.

The 'living process' concept requires us to view our tools in a new light. Our view of ATE systems changes from *testers* to *process monitors*. Testers are used to sort boards into good or bad piles. Process monitors do this as well, but they also collect data about the process upstream from that point. These data contain both direct and inferential information. An example of direct information might be

Resistor R17 failed:
Measured 862 ohms
Expected 1 Kohms +/- 5%

If this component fails over multiple tested units, then we can infer that something in our process has changed. It could be simple, like a misload of an 850 ohm resistor or maybe the resistor vendor is sending resistors produced by a new process that does not meet the required specifications. In a living process, we study the evidence until a cause is found, and then we adapt. If it was a misload, how did it occur? If a vendor is failing to deliver contracted parts, work with the vendor (or change vendors). In the end, the problem is resolved with the process improved as a result. As improvements continue to be made, new problems tend to decrease in frequency. Sometimes, problems that escape our ATE monitor will be brought to light. Then it becomes possible to improve the test program to prevent recurrence. We now have a process capable of modifying itself, with each evolution an optimization over previous performance.

This view of living process has interesting implications for CAE systems. For example, a problem with the process is monitored over time by our ATE, and inference shows that the root of the problem is in the *design* of the circuit. We then may need to try out additional studies and potential solutions on the CAE system long after the original release of the product. Note that this release event was often called a 'freeze.' In the living process concept, the word freeze should disappear. Changes should be expected and the process infrastructure set up to deal comfortably with them. This does not mean that the original designer has to be found and brought back to fix the problem. Ideally, the original design database migrated from the design group into production, where it *lives on*. Problems are solved by using these data, which are then updated. Because we did not disturb the original designer, the development cycle for the next product is not interrupted. The designer is merely alerted so that a repetition will not occur.

1.5. Summary and References

Today, we are witnessing the redefinition of the product development cycle. The driving forces for this change are

1) Competition
   - Shorter Time to Market
   - Increasing Quality
   - Better Product Price/Performance Ratios
2) Emerging Tools
   - Personalized Computation Power
   - CAE Software/Accelerator Hardware
   - Design and Test Networking
   - ASIC Technology
3) Desires
   - Better Use of Human and Capital Resources
   - Integration of Design and Test
   - Shared Information
   - Parallel Development
   - Reusable Development
   - TQC and the Living Development Process

It is clear that CAE and ATE systems play key roles in this change and that there is synergism in their combination.

[Balliew 1984]	R.E. Balliew, "Automatic Test Program Generation for Digital Board Testing," *Hewlett–Packard Journal*[1], Vol. 35, No. 10, Oct. 1984.
[Crook 1979]	D.T. Crook, "Analog In–Circuit Component Measurements: Problems and Solutions," *Hewlett–Packard Journal*, Vol. 30, No. 3, March 1979.
[Feigenbaum 1983]	A.V. Feigenbaum, "*Total Quality Control*," 3rd Edition, McGraw–Hill, New York 1983.
[GenRad 1984]	*Introduction to In–Circuit Testing*, GenRad Inc., Concord, Mass., May 1984.
[Hall 1985]	R.W. Hall and J. Nakane, "*Comparison of Just–In-Time Production by American and Japanese Manufacturers,*" Association for Manufacturing Excellence, Elm Grove, Wisc., 1985.
[Harwood 1984]	V.R. Harwood, "Safeguarding Devices Against Stress Caused by In–circuit Testing," *Hewlett–Packard Journal*, Vol. 35, No. 10, Oct. 1984.

[1] The Hewlett–Packard Journal is available from:
Editor, Hewlett–Packard Journal
3200 Hillview Avenue
Palo Alto, CA 94304 U.S.A.

[Parker 1986] K.P. Parker, "Testability: Barriers to Acceptance," *IEEE Design and Test of Computers*, Vol. 3, No. 5, pp. 11–15, Oct. 1986.

[Peters 1984] T.J. Peters and R.H. Waterman, *In Search of Excellence: Lessons from America's Best Run Companies*, Warner Books, New York, 1984.

[West.Elec. 1956] Western Electric Co. *Statistical Quality Control Handbook*, Western Electric Co., Indianapolis, Ind., Sept. 1956.

[Williams 1983] T.W. Williams and K.P. Parker, "Design for Testability: A Survey," *Proceedings of the IEEE*, Vol. 71, No. 1, pp. 98–112, Jan. 1983.

INTEGRATING DESIGN AND TEST: USING CAE TOOLS FOR ATE PROGRAMMING

Chapter 2
A General Model of a Digital Tester

Chapter 2

A General Model of a Digital Tester

This chapter will define a model of a digital test system and a suitable vocabulary for the rest of the book. From now on, we will only consider those aspects of a system used for digital test, because most of the work done on CAE workstations and simulators addresses digital logic. Also, most ATE today is capable of automatic program generation for analog components if access and control of such components are good. Where it is not, analog tests are written manually on a case–by–case basis.

It is also important to note that we will consider digital testing in a generic sense with little need for differentiating due to packaging. This is to say, it matters little whether we are testing ASICs at the wafer stage or functionally testing an entire board. Our model of the testing process as far as CAE tools are concerned is virtually the same. From now on, 'digital testing' will be considered generally except where some differentiation is required.

2.1. A User's View

How should the user of a CAE system view an ATE system? Quite simply, if the CAE user is viewed as an entity that performs experiments upon a software model of the circuit under design, then the CAE user should view the ATE system as a machine that performs physical experiments upon the actual circuit. These experiments, performed by ATE, are termed 'tests.' Some very large differences exist between these software based 'experiments' and hardware (ATE) based 'tests.' The human user examines the calculated responses and uses judgment in the interpretation of the results. The software model is an approximation of the real circuit. Again, judgment is used to filter responses for aberrations that are attributes of the modeling process rather than true characteristics of the circuit. The model may be changed as a result of some experiments or the experiments may be changed. The CAE user refines the design of the circuit by using these results and *verifies* that the design meets all design goals. This process of *design verification* is a highly creative, judgment–driven, human, and imperfect process. For all but the most trivial designs, it is quite impossible to *prove* that a design meets the desired objectives in all respects. Indeed, the stated objectives are necessarily incomplete because of all the simplifying assumptions used to generate them.

The ATE system, on the other hand, is performing tests on a physical device. Here, real volts, amperes, and nanoseconds are involved with all the attendant laws of physics. These laws dictate limits that may not have been observed during simulation. Other limits exist for the tester defined by its basic capabilities. The ATE system performs simple judgments, usually PASS/FAIL in nature, as programmed. It must be supplied with inputs to present to the device under test, and it must be told what to expect from the device as a result. If the device does not respond as expected, this is indicated as merely PASS/FAIL information (one bit) or in more detail (many bits). Such detail may include failing signals, their states, which test(s) failed, etc. In general, the more detailed responses are called diagnoses, even though they often are only a description of the symptoms of a problem and not a true diagnosis. In some cases, particularly for functional board test, additional data may be gathered and correlated with a database created from simulation. The results of this effort can sometimes give much improved resolution of the true cause of the problem.

A perfect ATE system, executing the perfect test, would give a precise, true diagnosis for any conceivable problem. In reality, a collection of symptoms (a fault syndrome) is gathered. A true diagnosis comes from analysis, again a human endeavor. Often, at IC test for example, a detailed diagnosis is not desired. We just want to sort good from bad. However, with the growing emphasis on quality improvements comes an increasing desire to know not just *which* are bad, but also *what* is bad about them, so that we can determine *why* they are bad and take remedial action. CAE systems are in an excellent position to provide the necessary data for this.

Figure 2.1 presents a general block diagram of a digital test system. Again, for the purpose of this discussion, it doesn't matter if the tester is an IC tester or a board tester. All such systems are made up of a sequence and control subsystem which controls a parallel collection of driver/comparator (D/C) units, often called 'tester channels.' There may be many hundreds of these units all running in parallel. Briefly, a D/C unit consists of a signal driver to supply circuit stimulus and a comparator that can be programmed for the expected circuit responses. In more complicated systems, each D/C unit contains its own supporting memory where stimulus and response data for many tests are stored. Then, under direction of the sequencer, all the units execute, at high speed and in parallel, a serial collection of stimulus/response experiments.

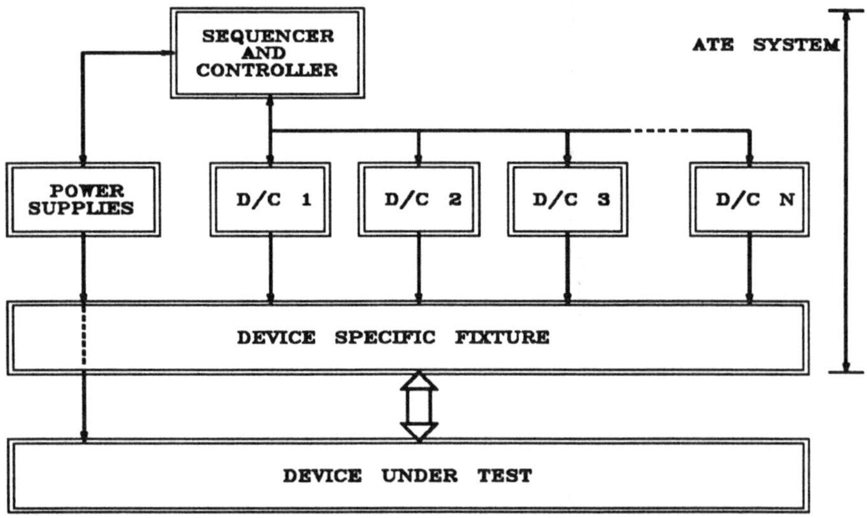

Figure 2.1

The outputs of the D/C units must be electrically connected to the device under test. To make a tester capable of testing numerous different circuits, an interfacing scheme is interposed between the D/C units and the device under test. The tester has a standardized physical interface onto which a device specific fixture is connected. The fixture is designed for quick connection and removal. The device to be tested is then mated with the fixture. The design of the fixture is influenced by various considerations such as signal integrity (discussed later) and whether or not mechanical handlers will be used to test entire batches of devices. IC test is often an extreme case for both handling and signal quality. An IC test fixture can cost $20,000, have 100MHz bandwidth, and interface to handlers that index devices in tenths of seconds. The entire test, comprising millions of stimulus/response experiments, may last only seconds.

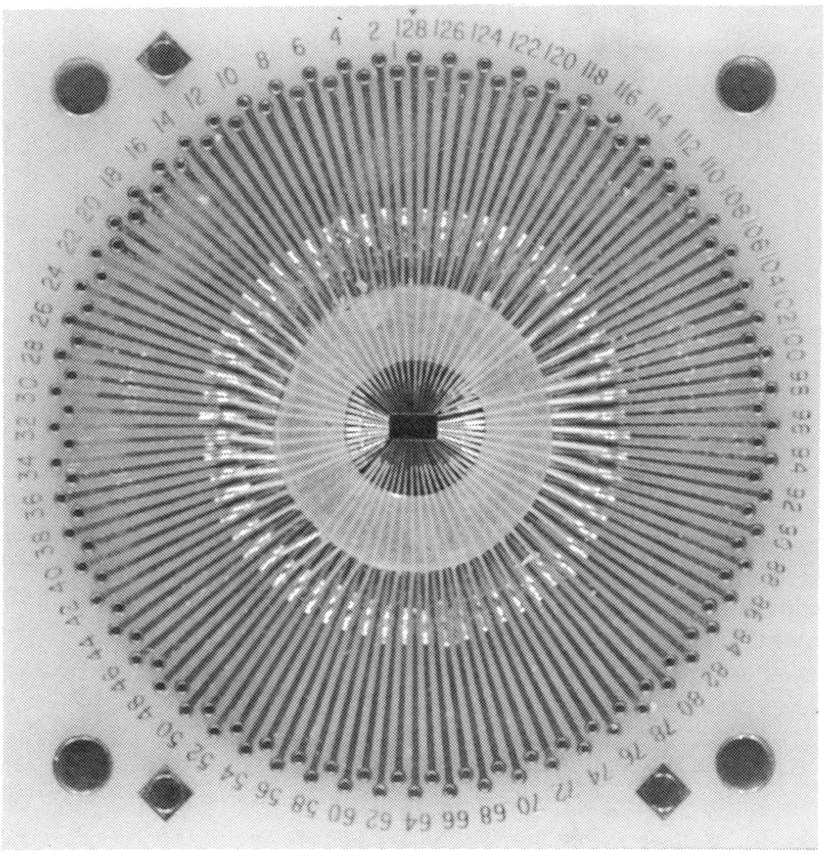

Photograph of an IC wafer test probe fixture magnified 1.5 times. This fixture is the final interface between the IC tester electronics and the device under test. In this case, the IC is still part of a wafer, together with hundreds of other copies of the IC. During wafer test, the probe fixture is positioned over an IC and moved only in the Z axis. The wafer is indexed in the X and Y axis to bring each copy of the IC into position for test. This probe card contains 64 contact needles in a 2x4 mm rectangle.

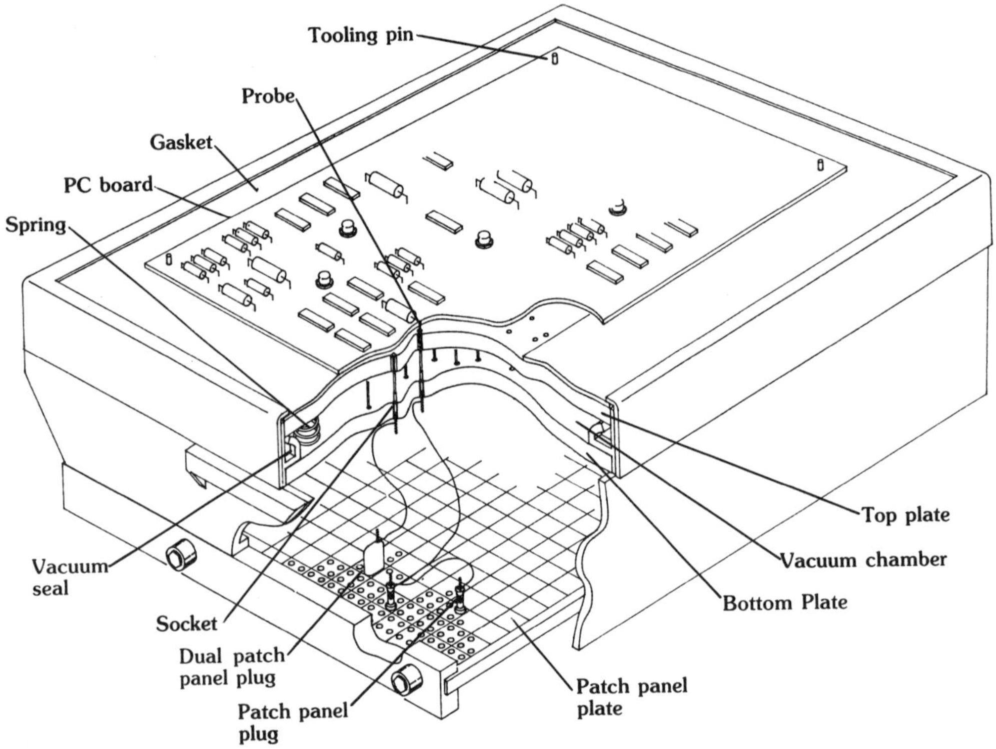

Cutaway drawing of an in-circuit test fixture. A vacuum chamber is formed between the top and bottom plates that allow a few millimeters of travel under vacuum, establishing nodal contact between the nails and the bottom of the printed circuit board. Each spring loaded nail may require 50 to 150 grams of compression force. If a large number of nails exist (for example, 1000) then more force may be required than can be supplied by the vacuum. The vacuum may be augmented or replaced by a mechanical compression scheme.

Designers in the past have lacked an appreciation of the problems of testing, undoubtedly caused by their isolation from the test department. There has been a tendency to look down on testing as some form of lower technology. It certainly is true that testers for tomorrow's technology must be fabricated from today's components. A closer examination shows that a fairly 'mundane' 5–MHz tester with 200 active channels is actually a massively parallel, programmable, arbitrary waveform generator/monitor with a data rate of

4 bits/channel X 5 MHz X 200 Channels = 4 Gigabits/sec (!)

(There are 4 bits/channel to specify data and control on both the driver and the receiver portion of the channel.) The awesome implication is that each second's worth of testing requires the preparation and manipulation of 500 Mbytes of test information. No wonder CAE tools are desperately needed. We might also wonder how testers can do this with only a few kbytes of memory behind each channel. Perhaps there is more technology in this mundane tester than meets the eye.

2.2. General Terminology

This section serves as a glossary of terms related to tester hardware and the data used to drive it.

2.2.1. Driver

A tester contains a set of parallel drivers used to stimulate a circuit under test. For the stimulus side of a stimulus/response experiment, all drivers are instructed in parallel to drive preprogrammed 'data,' usually fetched from memory dedicated to each driver. This is the beginning of a 'test cycle.' (In most testers today, the drivers are combined with other replicated resources such as comparators.) Drivers are three-state devices: driving high, driving low, and not driving (high impedance). The high and low states are defined by reference voltages. In some cases they are fixed (e.g., standard TTL levels) or they may be programmable. See Figure 2.2.

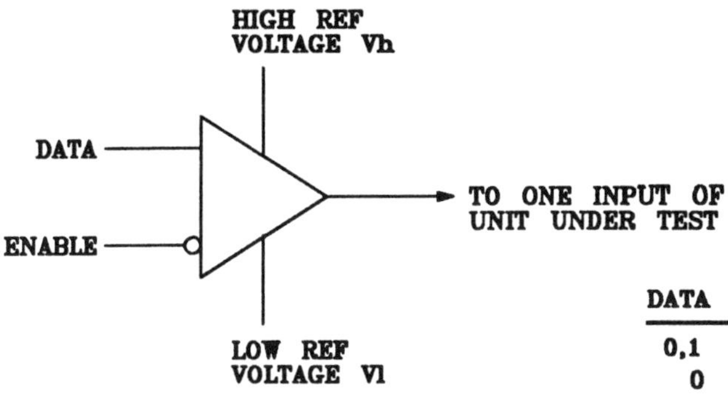

DATA	ENABLE	OUT
0,1	1	Hi-Z
0	0	Vl
1	0	Vh

Figure 2.2

A driver has a finite steady state current capacity for which it can achieve the reference levels. This capacity is usually high enough to satisfy the input requirements of the circuit under test. However, when making a transition from one state to another, it is likely that the transition time will be influenced by the amount of current loading presented by the device under test. Most drivers are designed to present a minimum of leakage and capacitive loading when in their high impedance state. It is important to note that the inevitable fixturing between the driver and the device under test will create additional loading effects, which must be considered at higher operating frequencies.

Some drivers have additional features. Some, for example, have selectable rise/fall times. Some have the ability to 'format' data. This is worth some discussion. See Figures 2.3a, 2.3b, and 2.3c.

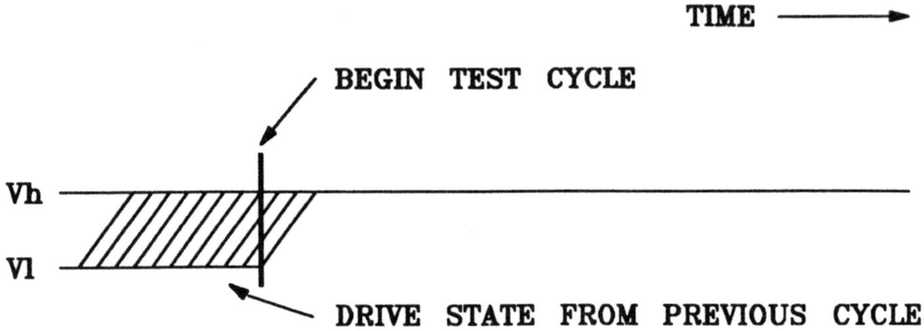

Figure 2.3a

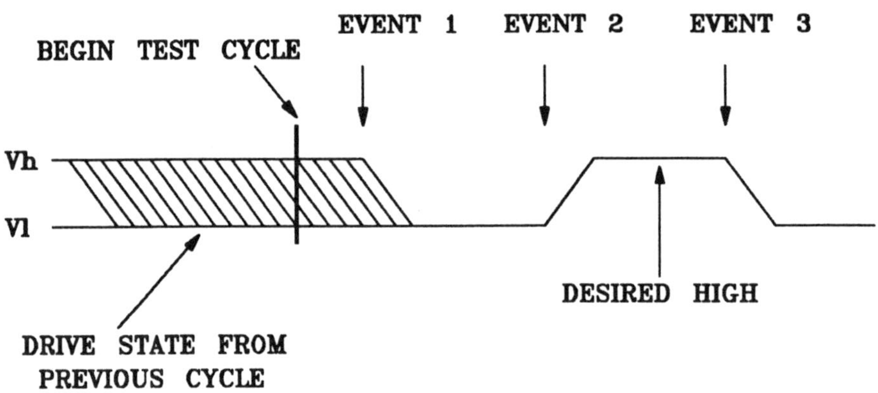

Figure 2.3b

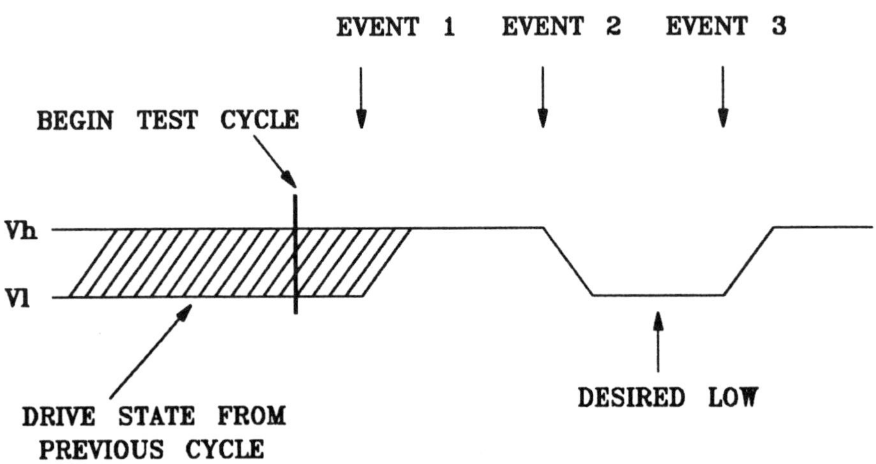

Figure 2.3c

Figure 2.3a shows a simple driver, driving high for a tester cycle. At the beginning of the cycle, the driver is in one of its three states and moves to the high state (Vh). Once at Vh, no other action occurs.

Figure 2.3b shows a more complicated driver also 'driving high.' This is a formatted driver. It 'modulates' data with a predefined wave shape. This example shows a 'surround–by–complement' (SBC) wave shape. Here, at the beginning of the period, the driver is in one of its three states as in the simple case. However, the driver waits until 'event 1' to begin driving *low*, the complement of high, which is the intended value to be produced. At 'event 2,' the driver drives high, the intended value, followed by another drive low at 'event 3.' In an exactly complementary fashion, the driver could drive low with an 'SBC' format, as shown in Figure 2.3c. We call these two waveforms SBC–1 and SBC–0. With these two waveforms, there are three edges possible per test cycle, but one bit of data is actually used to define which waveform occurs. The driver has been preconfigured with information about the modulating waveform. This information includes the number of 'events,' where in time they occur, and how to modulate the data. The most complicated drivers not only format data, but allow the storage of multiple formats that can be selected 'on–the–fly' during the execution of many test cycles. Needless to say, these are also the most expensive drivers. Other formats often seen are 'return–to–zero' (RZ), 'return–to–one' (RO), 'non–return–to–zero' (NRZ), and 'surround–by–Hi–Z' (SBZ).

Many board testers utilize the simple drivers seen in Figure 2.3a. In more advanced board testers and in virtually all IC testers, some amount of formatting capability (Figures 2.3b and 2.3c) is available on some or all drivers. Why? What are the implications for users of CAE?

Formatting of driver waveforms began on IC testers. In essence, the amount of 1s and 0s data stayed constant while the number of edges per data bit increased. This is done to provide *stress* on input timing specifications; that is, on the AC parameters of the device under test. For example, consider a simple edge triggered D flip–flop shown in Figure 2.4. To load a '1' into it, we simply put a 1 on the D line, and some time later place a rising edge on the clock line. Figure 2.4a shows this simple and nonstressful, stimulus using two simple drivers. Figure 2.4b shows the D driver changed to a 'stressful' surround–by–complement format for the same test. In this case, we are simultaneously stressing two timing specifications, the data setup and hold times, as well as loading a 1 into the flip–flop. As far as logical data are concerned, the two tests are identical. The formatting capability is used to stress AC parameters T–setup and T–hold.

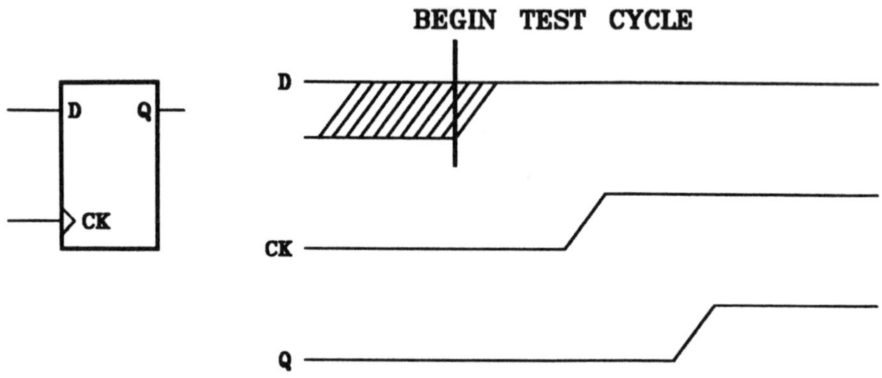

Figure 2.4a

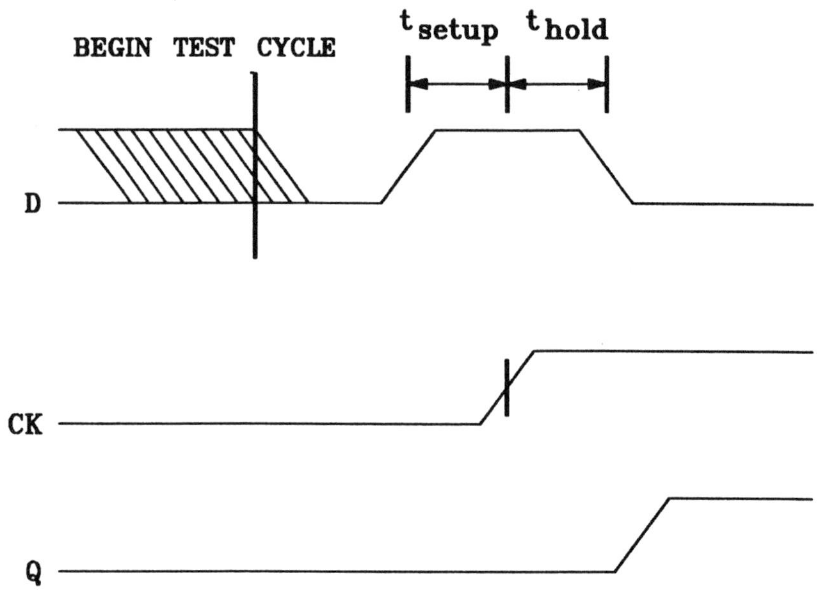

Figure 2.4b

Continuing with Figure 2.4, if we were to simulate either the relaxed or the stressful waveforms on the CAE system, the flip–flop should load a '1.' It could be argued that for simplicity's sake, we should use the simple waveforms on the CAE system and let the test engineers add the formatting 'stress.' Indeed, this is often done. It is not always feasible to take 'stressful' waveforms from a CAE system and make the tester driver resources replicate them. The CAE user preparing tester data should form a consensus with the test engineers about the issues of formatting and whether formatted waveforms should be used during simulation or added later.

There are a multitude of formats that fancy testers can generate. In the world of board test, formatting is sometimes available. Here, formatting can be used to stress some timing parameters, but there is a more interesting use.

By utilizing a small number of formatted drivers, many 'typical' board tests can be developed with lower overall driver frequencies. This is to say, only a few signals on a board may need to operate at elevated frequencies while the majority are slower. A good example of this is a simple RAM board. Such a board has an address bus, data bus, and a few control wires typically labeled READ/WRITE, ENABLE, SELECT, etc. Thinking in terms of read or write cycles, one finds that the address and data can be presented as single edges per cycle while the control lines may have two or three within a cycle. Furthermore, the control functions are not naturally thought of as 1s or 0s but rather as READ, WRITE, READ–MODIFY–WRITE functions, etc. By using formatted (multiple edge) drivers for the control lines and simple drivers for the address and data buses, tests can be very naturally stated as read and write cycles. Thus, there are fewer overall tester cycles that must be specified and stored in channel memory because the multiple edge control functions compact the test.

2.2.2. Receiver/Comparator

Another parallel and replicated resource in a typical tester is the receiver/comparator or, simply, the comparator. These are connected to the outputs of the circuit under test and allow us to observe its behavior, with respect to what we *expect* it to do: its *expected response*. See Figure 2.5 for a conceptual model of a comparator.

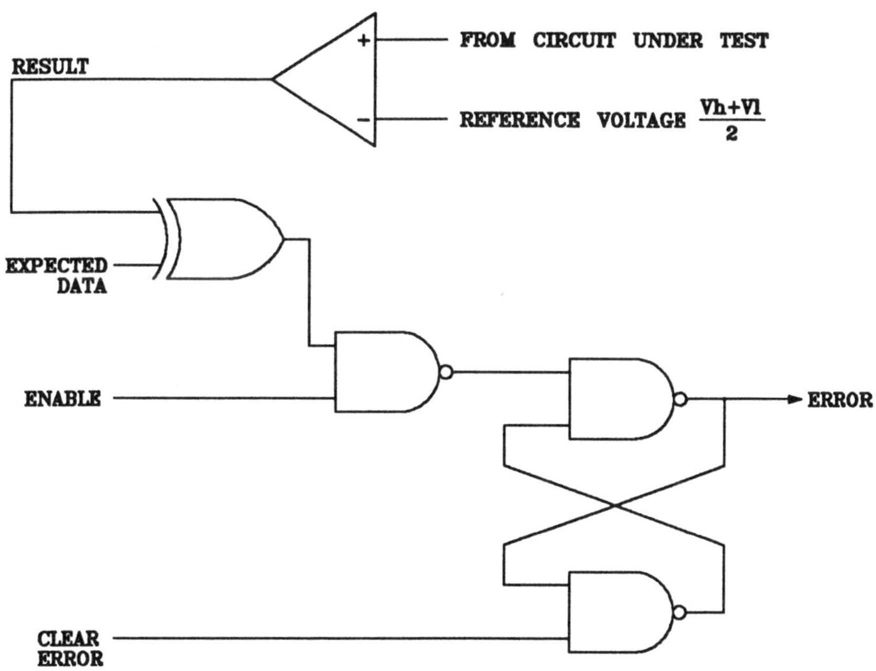

Figure 2.5

This simple circuit receives the output from the circuit under test and buffers it from the rest of the comparator to minimize loading. The single REFERENCE VOLTAGE is used to generate the signal RESULT, which is 1 if the test output is above the reference and 0 otherwise. This RESULT is compared (exclusive–or) with the data we expect to see. Then, the comparison is masked with an ENABLE. Controlling ENABLE allows us to choose when to look at the circuit response and for how long. Some testers use a window scheme where ENABLE is held high for a length of time and the tested output is expected to be good the entire time. Others use a sampling scheme, which is like an infinitesimal window that can be positioned at will. The result is latched by the cross–coupled NANDs, which are armed before the test cycle starts. The signal called ERROR is 1 if the tested output is not as expected any time during the sample window.

Digital logic, for enhanced immunity to noise, has a gap between acceptable voltages for low and high logic values. A more sophisticated comparator, shown in Figure 2.6, has two references and comparators that determine if the signal is above high (AH) or below low (BL) voltage references Vh and Vl. A signal from the circuit under test that falls in the noise gap will be judged to fail, since both AH and BL will be 0. This is known as dual threshold comparison.

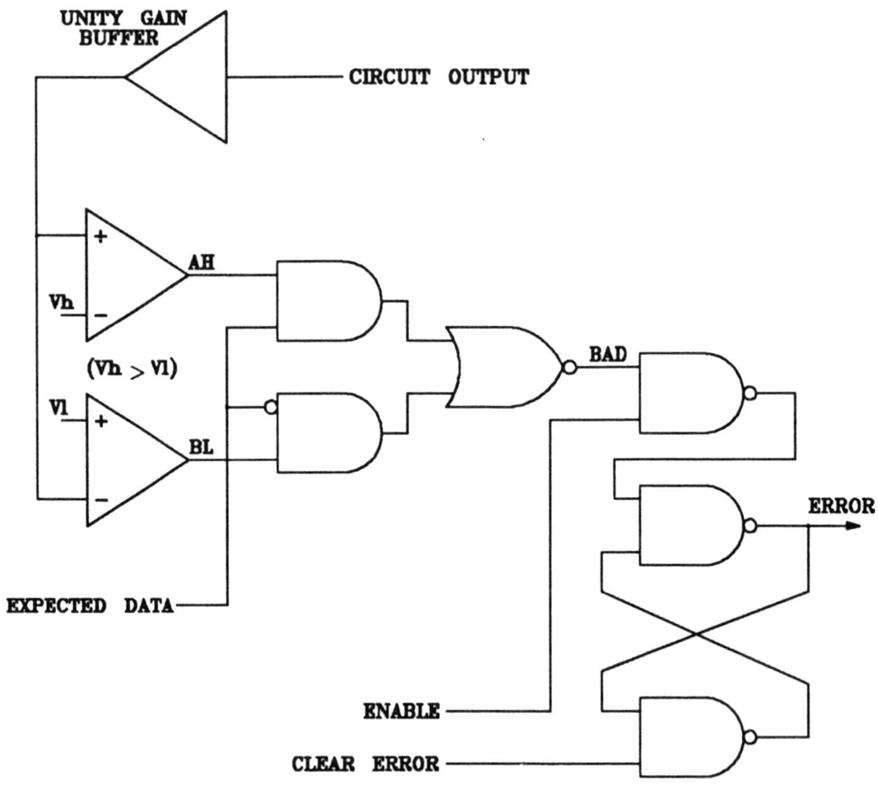

Figure 2.6

This comparator first buffers the incoming signal to reduce loading. The signal is then fed to two threshold comparators referenced to Vh and Vl respectively, producing the 'above high' (AH) and 'below low' (BL) signals. The EXPECTED DATA signal is used to choose which of these signals is presented to the error latch during the sample window as before.

The main parameters that a tester controls for comparators are the threshold voltage(s), the expected data, and the characteristics of the comparison window. Thresholds are fixed before testing begins. The expected data must change from test to test to correspond to the operation of the circuit under test.

Simple testers use sampled comparison with a fixed delay into the test cycle as in Figure 2.7a. After the drivers put out data, time is allowed for the circuit under test to settle. Then the response is sampled. Figure 2.7b shows a very complex tester driving SBC type data to the circuit and opening comparison windows at two different times. Such a test could conceivably be used to check data setup, data hold, and min/max propagation delay on the D flip–flop of Figure 2.4.

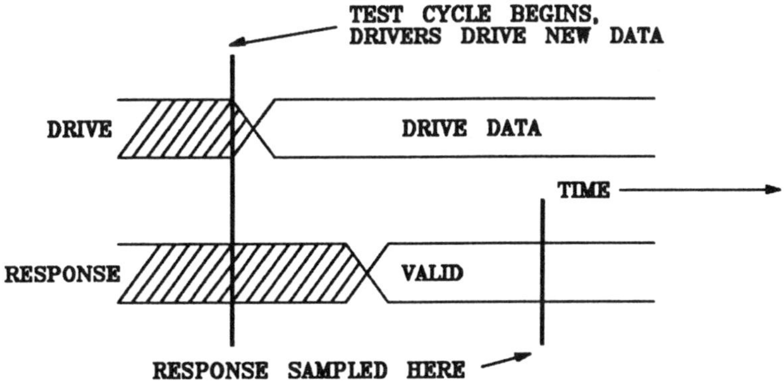

Figure 2.7a

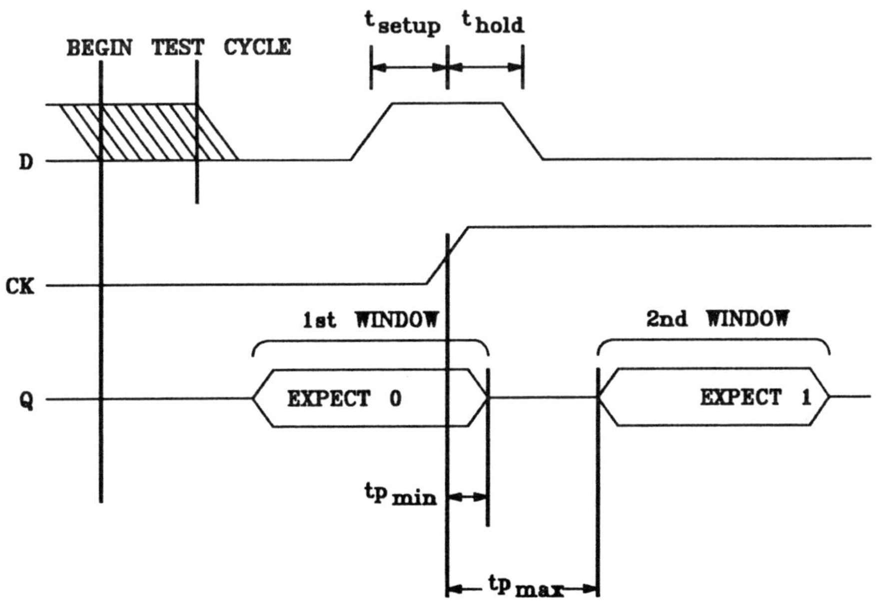

Figure 2.7b

Notice that two bits of compare data and four placements of edges on ENABLE are required for the comparison. This level of complication requires a highly sophisticated tester with extreme accuracies. Typically these are million dollar high-end IC testers used to characterize ICs.

On some testers, the comparators have an additional task. Here, they are paired with a driver in each channel. If the channel driver is in use to provide input to the device under test and the comparator would be otherwise idle, it is connected to the driver. It is programmed to expect the same data that the driver is supposedly driving. If the driver cannot achieve either logic level because of a short, for example, the comparator will flag an error. The occurrence of this error can stop the test before damage occurs to either the device or the tester, although robustly designed testers should easily survive simple problems like this. Testers that use this 'comparator follower' scheme do so automatically; the test programmer does not have to set it up explicitly.

The user of CAE will find it easiest to consider the simple model of the comparator. A single stimulus/response experiment consists of drive events that cause the circuit under test to react. At some point, a resultant state should be reached. Then, the comparators can be strobed (sampled) for the expected result, which the simulator can provide. This very simple model can be modified to put several drive events before the comparison is done, mimicking formatted drivers. One must ask, however, if this is necessary. Again, the example used in Figure 2.4 shows how CAE can be used to prepare the simple case and the test engineer can then augment the waveforms to test additional parameters.

2.2.3. Signature Registers

Some testers offer the ability to compress the response from the circuit under test into a compact signature. Signature compression, using cyclic redundant counters (CRCs) [Nadig 1977, White 1978, Williams 1983], can be used to reduce a long stream of response bits from the circuit under test into a 16 bit 'signature.' Here, rather than comparing each received response to an expected response, all responses are accumulated in the CRC and then the resulting signature is compared to an expected signature. This expected signature can be 'learned' from a known good circuit. An important characteristic of CRCs is that all bits must be accumulated before the comparison with a good signature is made. It is not possible to sense in the middle of the response stream that an incorrect response has occurred. The stored response comparison scheme seen earlier is able to indicate which response(s) is (are) bad immediately.

2.2.4. Active Loads

Active loads are another replicated parallel resource found on some testers, usually IC testers. They are mentioned here in passing. In digital testing, they are used to approximate loading conditions on circuit outputs in an effort to mimic real-world loading or specified load circuits often found in IC data books. We must be aware of them if creating a test on a CAE system capable of taking loading into account. Loading will significantly affect timing and voltage threshold levels. If we are trying to build a test that pushes the limits of the device under test and the ATE system, it is likely that loading conditions generated by active loads as well as those by tester and fixture parasitics will necessitate some careful debugging and tweaking of the test.

2.2.5. *Vector*

A 'vector,' in the world of ATE, is the specification of a set of parallel events performed by drivers and comparators. A common synonym is 'pattern.' An example of a simple vector may look like this:

```
"0011011zz0100zzHLHxxLLHxx"
```

Here, 25 tester channels are specified. Implicit in this example is the test cycle period. The characters '0,' '1,' and 'z' are associated with drivers and specify the action the driver is performing. (If formatting is occurring, that too is implicit.) The characters 'H,' 'L,' and 'x' are associated with comparators and list the data the comparator is expecting. Note that 'x' means the comparator will not be enabled at all during the test cycle. This occurs when we are not attempting to judge the results of a tested output and is often called the 'don't care' state. Indeed we may not care, but it also may happen that we 'don't know' the circuit output at this time. This will occur when the circuit (or part of it) is uninitialized and we (or a simulator) cannot predict an output value. Thus we must disable the comparison.

A set of vectors, also called a pattern set, might look like this:

```
"0011011zz0100zzHLHxxLLHxx"
"1111111101110OOHHHHHHHHHH"
"1011011HH0100LHHHHHHHHLHH"
"00110110L0100H1LLLLLLLHHL"
"0011011L101001LLLLLLLLHHL"
```

Here, five test cycles have been specified to be executed serially by 25 parallel tester channels. One notes that four of the channels have the characters 'z,' '0,' '1,' 'H,' and 'L' specified at various times. This indicates that the pin under test is driven at some times and compared at others. This is how a test for bidirectional pins might look. When preparing a test for a circuit that has bidirectional pins, one must take care not to drive them when they themselves are trying to drive, or contention will result, perhaps even damaging the circuit.

2.2.6. *Sequencer*

The sequencer in an ATE system causes the parallel collection of drivers and comparators to 'execute' vectors. It governs the period of each test cycle. In simple testers, the vector data for each channel are stored in a dedicated RAM, and all the sequencer does is send out the addresses of the vectors in ascending order. In more complex sequencers, addresses are still broadcast to the tester channels, but they may be random in nature as higher level concepts such as loops, subroutines, and data compression are used in the development of tests. In some testers today, vectors are written in a user–oriented language, and a compiler generates the information used by the sequencer to execute the test properly.

Some sequencers also have 'algorithmic' sequencing capabilities. They may be able to synthesize 'counters' or 'graycodes' as well as some of the more complex RAM test algorithms [Sohl 1977] such as MARCH or GALPAT.

The sequencer may also look at the results (the 'OR') of the ERROR signals from all the comparators. If one or more comparators is signaling an error, the sequencer may halt and give a FAIL indication. There are two reasons for halting immediately. First, in practical testing situations, little or no useful information will be gathered by continuing the test, due to the cumulative effects of faults on the internal state of the device under test. Second, it is conceivable that the tester drivers may be (or begin) 'fighting' with bidirectional device pins, causing unhealthy current loads, because the circuit is no longer listening when the tester is talking.

2.2.7. Digital Test

A digital test is a collection of properly sequenced vectors. Their execution will exercise the circuit under test and any malfunctions within it will show up as one or more comparators giving an ERROR signal during one or more vectors. A digital test may have a preamble of vectors called a homing sequence. Homing sequences will be covered shortly. If a homing sequence is not required, then a synchronizing sequence is embedded in the beginning of the test. This is covered next.

2.2.8. Synchronizing Sequences

For a digital test to be meaningful, the circuit under test must be brought to a known state. When first powered up, the circuit's memory elements (latches, flip–flops, RAMs, registers, etc.) are in unknown states. Many circuits have a 'hard reset' that can be used to create a completely defined initial state with the assertion of a single signal. For others, a sequence of inputs exists that will also result in an initial state, without need to examine the outputs and change the inputs along the way. Such vector sequences of length 1 or more that are applied without regard to circuit outputs, are called synchronizing sequences. (These will be seen to be special cases of homing sequences in that we make no changes in inputs based on the observed outputs.) As a result, testing of known outputs can proceed during synchronization. Those outputs that are unknown at earlier stages of the sequence must be ignored by the comparators.

2.2.9. Homing Sequences

A homing sequence consists of stimuli that drive the circuit from an unknown state toward a recognizable state from which the actual test proceeds. During the homing process, the comparators are used to look at the circuit output and indicate when the desired state has been reached. The stimuli may need to be repeated or altered depending on what the comparators observe during the sequence. While this is being done, the ERROR indicator cannot be interpreted as an error. Testing begins when the desired state is achieved. It is possible that a circuit failure will prevent the homing sequence from reaching the desired state. This condition could lead to an infinite loop, and thus the homing process usually has a time out limit. The normal test cannot be run meaningfully if the proper initial state cannot be achieved.

The need for a homing sequence is a property of the design of the circuit under test and is not an indication of some weakness in the tester. As an example, consider the case of a simple circuit, a 74161 4–bit binary counter (Figure 2.8) with parallel load and clear.

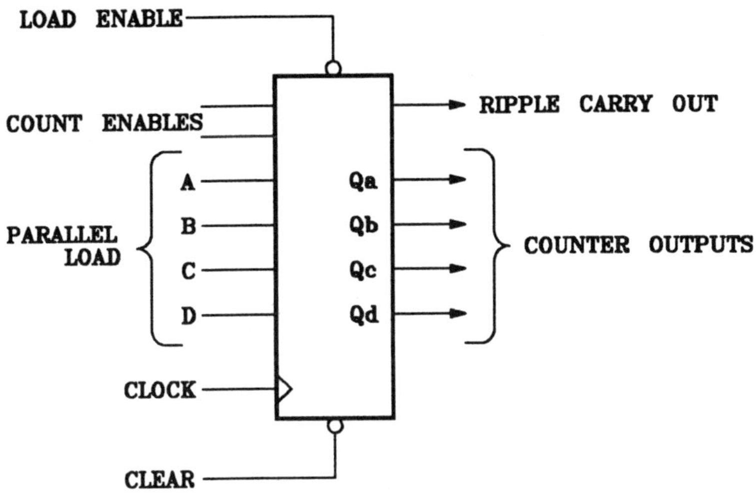

Figure 2.8

A one–vector synchronizing sequence can be done by asserting clear. The counter goes from the unknown state to complete initialization and the outputs can be checked by the comparators near the end of the vector. Now, if the clear line is unavailable (e.g., tied high), a two–vector synchronizing sequence can be implemented as follows: First apply 0s to the four parallel load inputs and drive the LOAD line low. This is vector 1. Then, on the second vector, apply a rising edge to the CLOCK line. At the end of vector 1, the state of the counter is still unknown, so the comparators cannot be enabled. However, the counter is completely initialized after vector 2. Finally, consider the case where both the clear function and the parallel load function have been eliminated (for example, LOAD and CLEAR both tied high to disable them). Now, no synchronizing sequence exists. However, a homing sequence can be created. The counter will start in some unknown state. We use a comparator to look at the RIPPLE CARRY output. We know that if we clock the counter between 0 and 15 times, this output will go high. When this happens, one more clock will place the counter in the same state as a CLEAR pulse would have. Thus, our homing procedure consists of looking at RIPPLE CARRY and applying CLOCKs, up to 15 times. If after 15 CLOCKs we have not seen RIPPLE CARRY go high, we can say the homing process has failed. This is not a very good diagnosis. In general, if we have the choice, synchronizing sequences are preferred; the shorter, the better.

If a circuit requires a homing sequence, this presents difficulties covered in Chapter 5 to the user of a CAE system. A valuable general treatment of the subject of initialization of automata is given in Chapter 3 of [Hennie 1968].

2.2.10. Test Plan

A test plan, also called a test program, is a complete set of critically ordered activities used to test a circuit. The test plan controls the basic operation of the entire ATE system, plus, it may be in charge of collecting test data utilized later for process control.

A test plan for a typical circuit board containing a mixture of digital and analog components might proceed as follows: First, unpowered tests are performed. That is, tests are performed upon the board before supplying potentially dangerous voltage and current to it. These tests utilize low level signals, well below those required to switch on transistor junctions. They look for short circuit paths and are also used to perform impedance measurements on analog components. When these tests pass, you are reasonably sure that the application of power will not damage the board. Then power is applied and checked for nornal operating limits after an appropriate settling interval. Once the board is 'live,' then testing may begin on digital components or collections thereof. Any failures during these tests are recorded. When completed, power is removed, including the bleeding of stored charges. A record of the test is made, perhaps physically on a strip of paper or in a database associated with a board identification code.

Test plans at a given level of a production process such as chip package test or board test have a common structure that varies little. Thus, most sophisticated test system software will automatically create a test plan. This may be confused sometimes with 'automatic test generation' which will be examined in Chapter 3. While it is not difficult to create the test plan, its component tests, such as analog tests or digital tests, can be quite challenging.

2.2.11. Diagnosis

Testing answers a basic question: Is this circuit good? The answer is binary – yes or no. A diagnosis is the addition of more information based on the observed behavior of the device under test. This information serves as a prediction of the true problem with the circuit and, as such, differs from the meaning of 'diagnosis' in the medical world. In the world of ATE, a diagnosis must be treated with care. The true problem with a circuit may only be found after additional testing and experimentation. It is not uncommon for the 'diagnosis' to be misleading or blatantly incorrect. At least the single bit, good/bad, is usually reliable.

Before becoming overly critical of this liberal usage of the word 'diagnosis,' think of the difficulty of diagnosing lung cancer by using only a thermometer and tongue depresser. At best one could get an 'initial' diagnosis. By using 'exploratory' surgery or X Rays, one gains new information toward the true diagnosis. Analagously, ATE systems often yield an initial opinion of the problem. When monitoring a process with ATE, it is important not only to record this information but also to record the results of future experimentation leading to the true diagnosis. In this way, it may be possible to improve diagnostic resolution and accuracy in the future.

2.3. Digital Functional Test

A digital circuit board (or IC) can be stimulated and observed from its edge connector (or bonding pads). This is called a functional test. Here, a major piece of digital circuitry is being tested with limited control and observability; that is, we aren't allowed direct access to internal circuitry. However, the circuit is being operated in a fashion similar to its ultimate operating environment. Indeed, the closer we can get to recreating the home environment of the circuit during testing, the better one seems to feel.

The types of failures that may be discovered have a nondigital flavor, such as inadequate timing margins, lack of current drive, or excessive leakage on a bus. It is possible to run at or close to the rated data rates of the circuit. There are no loading effects injected inside a circuit board, allowing a more realistic measure of how well the ICs play together.

One criticism of functional testing needs to be mentioned. While it is widely known that test preparation for functional testing is *extremely* difficult and that it is hard to measure the effectiveness of such a test, the real problem with it is strategic. It is often the case that the fault spectrum being addressed contains a class of design errors. For example, the design may contain marginal timing relationships and for some mix of components, a percentage will line up in a way that violates a timing relationship. A high speed functional test *may* detect this problem. It is another story altogether to actually isolate this problem. One finds a very expensive tester is being used to painstakingly track down a design flaw. A much better strategy is to find such problems during the design stage by using the appropriate CAE design tools. The proper use of a functional test is to validate *known* critical timing relationships, not to shoot in the dark for unknown problems.

2.3.1. Edge Connector Test

As was mentioned, edge connector functional testing connects to the unit under test only at that point where the unit is designed to interface with the outside world. This usually provides a good match between the loading the unit can drive and the tester. Another advantage is that boards with common edge connectors can share fixturing.

With respect to test preparation, the entire unit must be treated as a whole. Thus, a board containing several hundred thousand gates must be considered as a single entity. This is a difficult proposition for simulators and other workstation tools that are always lagging increases in circuit density. The other issue is that because of the limited visibility into the circuit, it is harder to isolate that one failing IC from the group. While fault dictionary techniques (described later) attempt to precalculate the symptoms that each fault may exhibit, these depend heavily on the fault assumptions used and may lead to poor fault isolation accuracy.

2.3.2. Cluster Test

Cluster functional testing is quite similar to digital functional testing except that in–circuit techniques (described soon) are used to isolate a cluster of ICs from the others around it. This means the physical limitations of in–circuit testing will have to apply, reducing some of the capability available with edge connector testing.

This form of test is coming into use where fixturing limitations, especially with respect to surface mounting technology, make complete nodal access an impracticality. The problems of failure isolation are the same as for functional board test, though somewhat less if some nodal access is still available within parts of the cluster.

2.3.3. Benefits

The benefit derived from functional testing is that the circuit has operated in a fashion fairly close to that of its end use. What this really means is that those failure mechanisms visible only during this type of testing have a *chance* of being detected.

2.3.4. Drawbacks

The drawbacks of functional test all relate to difficulty: difficulty in preparing the test, difficulty in executing the test (at the frequencies desired), and difficulty in reaching a proper diagnosis from the results. Test preparation difficulty is striking when compared to in–circuit test preparation. The time, effort, and skill required may be an order of magnitude higher. And yet, after this investment, the quality of the diagnoses may be significantly less. All of these difficulties have immediate economic impact and should be carefully considered.

2.3.5. Diagnosis with Dictionaries

We have already discussed the concept and limitations of 'diagnosis' as used in testing. When a test is run with the limited visibility afforded by functional testing, a failure has certain visible symptoms, collectively termed a 'fault syndrome.' One symptom is the first failing test (vector) number. The other is the output pin or pins that were in error at that time. Future tests (vectors) may also fail, but the effects become cumulative for sequential circuits and we simplify matters by only considering the first failing vector. Now, in the earlier years (see [Agrawal 1972]) it was realized that simulators were very useful for preparing tests for ATE and that they could simulate the behavior of a properly functioning circuit as well as that of a circuit containing a malfunction. Given a list of malfunctions (a fault list), one could simulate each by itself against a proposed set of test vectors. The simulator could record how the resulting circuit behavior differed from the normal behavior, in terms of which output pins would show an error versus a vector number. This gives a three dimensional database: the dimensions are (1) vector number, (2) failing output pin(s), and (3) the list of faults. This database is called a fault dictionary. When one executes the test against the circuit, one can correlate the fault syndrome (failing vector and pins) with this stored data and 'look up' the fault that caused the problem. This would give a good indication of the internal problem causing the externally visible syndrome.

In real practice, this approach has achieved limited success. The prime culprit is the assumption that the modeled faults are representative of those that will occur and that only one will occur at a time. Often, the look up effort comes up empty, leads one astray, or points to a large number of possible problems with no resolution between them. While not completely ineffective, dictionary diagnoses should be verified by some independent means. One such means is by guided probing, covered next.

2.3.6. Diagnosis with Guided Probes

Functional testing at either board or cluster level suffers from low diagnostic resolution. This problem can be approached by using a low capacitance guided probe (usually hand held) that gives the test a mobile response channel. The activity of internal circuit nodes can be measured and compared against expected data. The tester controller can look at a given node, decide if it is good or bad, and ask for new probe measurements at other nodes after examining the probable cause and effect relationships dictated by circuit topology. An excellent discussion of guided probing methodology is given in [Groves 1979], where complicating factors such as circuit feedback, inaccessible nodes, intermittent circuit behavior, and operator error are examined.

Guided probing has been very successful at enhancing the diagnostic resolution of fault dictionaries. Dictionary information is used to start the probing process at the indicated location of the fault and hence can be used to reduce the amount of probing necessary. However, if the dictionary prediction is wrong, the guided probe can still be used to backtrace from a failing receiver to an embedded failing IC.

Guided probing efficiency is adversely affected by certain logic structures. One example is in bused systems where a bus wire is faulty. Probes rely on logical (voltage) information, whereas current information (electron flow) is needed to separate the many contributors to a bus state. Another example is in asynchronous feedback loops where every element of the loop stabilizes to a bad value within one test period. The probe cannot utilize time information to break the cause and effect relationship down any more finely [Groves 1979].

One other key assumption is that a probe can gain access to the board. If the board is covered with an impervious coating, is dual sided, hybridized, or implemented with surface–mount technology, this assumption may break down.

Nevertheless, guided probe isolation has succeeded in the past. It requires node state data for every accessible node for every test pattern, but this volume of information is usually much smaller than a fault dictionary.

2.4. Digital In–Circuit Test

Digital in–circuit board test is accomplished with bed–of–nail access to all signals on the board. Any internal signal will either be driven or observed (received) at some time during a test. During drive, the normal driver for a node is overdriven (back–driven) and essentially overwhelmed so that control of the node belongs to the tester. This allows the tester to test an IC by overdriving all upstream devices while examining the IC outputs. The IC under test is tested as if it were independent of the rest of the board.

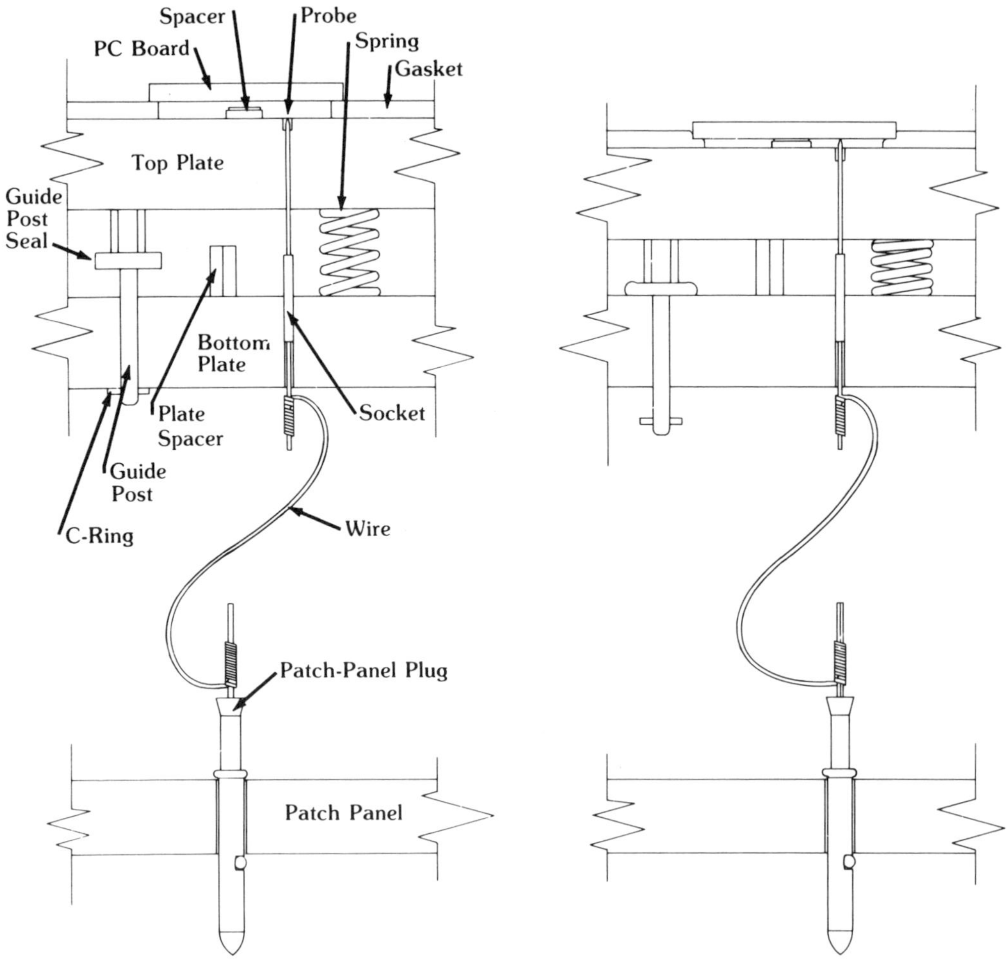

A cross section of a test fixture before and after vacuum is applied

2.4.1. Benefits

Each IC is tested, one by one, with those failing listed out. Thus, for boards with several failing ICs, only one pass on the test system finds them all. Fault diagnosis is improved because the high degree of nodal access prevents most interactions from other circuitry, although some are still possible in the form of short pulses that can 'sneak' past the overdriving levels. See the following section on restrictions.

Because of the high nodal access, it is possible to perform a non–powered shorts test (i.e., a continuity test using voltages below the turn–on threshold of typical logic families, typically 0.2 volt). Shorts (and opens) can be found between nodes before power is applied to the board. This protects against damage that power might cause and removes a failure class that is more difficult to isolate during digital testing.

One of the greatest benefits of the in–circuit approach is that tests for many common devices can be programmed once, in advance, and stored in a library. Since the device will be tested as if it were standing alone, the complication of developing test programs that is so difficult for functional testing is virtually eliminated. One exception is that some device inputs may be tied together or to fixed logical signals such as GROUND ('0'). However, knowing this is often done, one can build a series of library tests that take commonly found constraints into account. The test preparation software uses all tests in the library that do not violate the constraints.

2.4.2. Restrictions

There are some restrictions that may have to be addressed when doing digital in–circuit tests. They tend to be topologically induced and may invalidate the attempt to make stand–alone IC tests developed on a simulator work for in–circuit testing.

Pin Constraints: Some input pins to an IC may be hard–wired to ground or VCC (logic 0 or logic 1), which cannot be overdriven during testing. Another example of pin constraint is when an IC output is fed back into one of its own inputs. Then either the output cannot be observed because it is being overdriven to provide an input or the test must be written to take into account that the feedback exists. Yet another example is when two or more inputs are tied together. In each case, the original IC test is unconstrained and not valid in the constrained environment. The IC level test prepared by simulation will have to be redesigned and resimulated to take these constraints into account.

Testing in Bused Environments: When an IC to be tested drives a three–state bus, the bus driving output cannot be observed if some other member of the bus is active. Each member of the bus will have to be coaxed into quiesence by using additional tester channels to disable it. It is possible that the pattern set to be applied to the IC under test is not compatible with the necessary disabling requirements, because of board interconnects. This difficulty may not be obvious under casual observation. For example, an upstream device must be placed in the Hi–Z state and some disable vectors must exist in the library. Unfortunately, they work by placing a '0' on a signal named ENABLE, and this signal is also connected by board topology to a pin that is on the IC to be tested and cannot be '0' all the time. One will have to find some alternate way to control the upstream IC.

Failure to Overdrive: In some cases, overdriving some upstream outputs may not be feasible because of their strength (usually in only one polarity) or because parasitic oscillation occurs. When this is discovered, the test may need to be rewritten.

Feedback: The overdrive technique may be frustrated by short pulses that return to the device under test as a result of applying tests. For example, the IC being tested may be in a loop with several other ICs and a feedback path may be sensitized such that the effects of vectors propagate back to the overdriven outputs of the upsteam devices. These overdriven outputs will try to make transitions. The inevitable inductance of the path will allow a momentary spike, which can disrupt the test, to occur. When this problem occurs, one must commit extra tester channels to points in the feedback path to inhibit feedback propagation.

Device Damage: The user of a CAE system has the ability to generate huge volumes of vectors. However, upstream devices may not tolerate being overdriven for the lengths of time that these vectors will require. If this is the case, the vectors may have to be partitioned into stand–alone units, each with an appropriate initialization added on. It would be better to resist the temptation to generate large tests just because it is possible. Each vector applied *should have a purpose*.

Physical Access: It may be mechanically difficult to access the IC under test or some other IC that has to be controlled or disabled. If physical access is not assured, then one may have to resort to functional testing at the board or cluster level.

2.5. Required Data

In this section, we specify the type of data needed for testing. Not surprisingly, this is influenced by the test approach: whether in–circuit or functional, whether additional diagnostic action such as guided probing will be used, and whether we are testing boards or ICs. Here, we omit consideration of data not generated or manipulated by simulation. For example, power supply currents or output voltage levels will come from some other source.

2.5.1. IC Test

For IC test, the data needed for testing are usually limited to vector data and how the IC pins are ordered with respect to elements of the vector. For example, the signal named 'CLEAR' may be channel 1 of the vector, signal 'ADDR12' channel 37 of the vector, and so on. Since ICs are almost always considered as irrepairable units, support data for enhanced diagnosis, such as internal topology or states, is not generally required. This may change some day if internal probing of ICs via lasers or electron beams becomes practical. Then IC test will much more closely approximate the board test problem, although data so collected by probing will be used as process feedback and not for repair.

2.5.2. Functional Board Test with Probing

This activity requires several different kinds of data. First assume the ATE system, as part of its test preparation capability, has a library of IC information that identifies IC inputs, outputs, and bidirectional IC pins. This will be combined with board topology information from the CAE (or CAD) system to guide the probe. The CAE system can also provide the internal state information (i.e., the states that occur on points that can be probed within the circuit) to be expected by the probe during the test. This information could be learned from a known good board. As with the IC test above, the test vector information and the relationship of the edge connector pins to elements of the vectors must be supplied.

The topology information consists of logic signal names and their connectivity within the circuit, usually specified as pins on devices. A list of device names and the type of each device is also required. Some CAE systems may also have physical location information available, which could be used during probing to locate probe points. Most ATE today will operate without this, but it will be necessary for automated probing. Sometimes, simple naming conventions for signals and devices are used to make manual location of probe points simpler.

2.5.3. *Functional Board Test with Dictionaries*

This type of testing is very similar to IC testing, and, indeed is often done because access is not allowed to internal signals on the board. For example, boards designed for harsh environments may have thick conformal coatings applied to them that prevent in–circuit access and make probing difficult. Holes punched through the coating may need filling. Such a situation requires vectors and information about the edge connector as before. The dictionary data must also be supplied. Again, this is a three–dimensional database relating vectors to failed pins to simulated faults. See the discussion of dictionaries above.

2.5.4. *In–Circuit Testing*

When using CAE systems to develop in–circuit tests, the minimum amount of data needed is the set of vectors and the relationship of the elements of the vectors to the pins on the IC, plus board level topology. This includes interconnect and device data so that the necessary overdriving conditions can be set up. It is assumed here that the CAE system is used both for the design of individual parts (ASICs) and for the board as a whole.

While we are busy designing ASICs at our CAE system, there is some additional information we can collect that makes board level test preparation much easier. The in–circuit technique, particularly in bus–structured logic, can be susceptable to interference from ICs surrounding the particular IC (or cluster of ICs) currently under test. Thus, while a pattern set is required to test an IC, 'disable vectors' will often be needed when it is time to test another IC electrically connected to this same IC. Disable vectors are used to place an IC into a quiescent, nondriving, or safely overdrivable state so that testing of a topologically connected IC can commence without interference. When preparing tests for an IC, designers have the knowledge and tools available to take a few more moments and prepare a second set of disable vectors. This small investment will save a lot of trouble later.

2.5.5. Data Requirements Summary

The following table summarizes the data needed for different test situations.

	Vectors	Pin Info	Topology	Node States	Disable Vectors
IC Test	X	X			
Functional Board Test with Probing	X	X	X	X	
Functional Board Test with Dictionary	X	X			
In–Circuit Test	X	X	X		X

2.6. Fundamental Limitations

The next section explores some fundamental limitations to the capabilities of ATE systems. Some of these are induced by economic considerations in the architecture and construction of a given tester. Others are the direct result of the laws of physics, laws not necessarily binding on simulations. In either case, a user of CAE who is targeting an ATE system should be aware that care is needed in matching the capabilities of the two systems. In virtually every instance, it is the *CAE system* that must be constrained.

2.6.1. Channel Resources

The resources considered here are the replicated parallel D/C channels containing drivers and comparator/receivers. Active loads are not considered because of the many variations in their use and the fact that they are deliberately used to stress parametric characteristics of the circuit under test. The effects of their use must be considered on a case–by–case basis.

Node Count/Multiplexing: When planning to test a circuit on ATE, take a count of the number of inputs, ouputs, and bidirectional signals the circuit has. Make sure the ATE system will support this. Take care to note that many ATE systems (usually in–circuit testers) multiplex their channels to several physical circuit locations. Thus, ask how many channels can be supported *simultaneously*.

Some systems may have their resources divided into independent sets, such as independent drivers and receivers, while others offer hybrid capabilities and true support of bidirectional pins. Make sure the configuration you need exists in the system. It is possible that the system will have to be expanded, or, if this is not feasible, you may have to change your approach. For example, if you have a limit on receiver count, you may have to do a two–pass test. You need to consider the effects of this restriction before you invest in expensive simulation.

Driver Rise/Fall Time: Real drivers on ATE systems have non–zero rise and fall times. Furthermore, these transition times may be functions of the current loading presented by the circuit under test or by the upstream devices being overdriven. However, many simulators treat inputs to the circuit under simulation as having zero transition times. Therefore, narrow pulses can be simulated that may only be 'runts' (not full swing) if produced by a driver. (See also, minimum pulse widths, below.) One might try to order the timing of input events to the simulation to times shorter than the rise/fall times possible on the tester. During the rise time of a signal from a tester (a signal that appears to be a ramp) there is some uncertainty as to when the circuit input actually sees the transition. This uncertainty may cause the circuit to behave differently from the simulated prediction.

Some simulators will allow us to specify a rise or fall time on a signal; this can be modeled by 'simulating' the tester driver as part of the circuit we are studying. We can then inject signal rise/fall times, as the tester would produce them, into the simulation of the overall circuit. This trick can also be used to account for skews between drivers. See the discussion below on pin–to–pin skews. One thing is certain, if we are pushing hard on the timing limits of our circuit design and the ATE system, proper accounting for non–ideal waveforms will be important.

Drive to Receive Time: In many testers, there will be a limit or setup time requirement on receivers after a drive event has occurred. This will limit the degree to which we can measure signal propagation delays with the tester. Further, the uncertainty in time when a receiver begins its comparison with respect to a drive event may be important. This uncertainty may not be well characterized in the tester specifications, and the effects of pin–to–pin skew may be a contributing factor.

Pin–to–Pin Skew: The drive/receive channels of a tester are replicated units made up of slightly dissimilar components that are spatially distributed over some hundreds or thousands of square centimeters. Even with expensive calibration hardware on high–end systems, these effects cannot be completely eliminated. Thus, drivers operating in 'parallel' will not produce simultaneous events on the circuit imputs, even though a simulator is perfectly happy performing this physical impossibility on a model. There will be similar uncertainties about when parallel receivers begin their comparisons. In tests that are pushing the limits of our circuit and/or the ATE system, we must be careful to consider the implications of the various contributors to timing error. We may be failing good components erroneously.

The digest of papers from the International Test Conference for the last several years has had entire sessions devoted to timing accuracies and calibration issues. In particular, see [Sugamori 1981] where an analysis is given that shows how testers with claimed accuracies in the 1 nanosecond region will accumulate errors as high as 7 nanoseconds.

Minimum Pulse Width: The driver resources of a tester are able to produce a pulse on a circuit input. The minimum width of this pulse can be fairly short (e.g., less than 10 nanoseconds) on high speed formatted channels. On unformatted channels, the minimum pulse width may be directly related to the speed of the sequencer since it may take two tester cycles to create the pulse. If this is the case, then it may also be true that the width of the pulse is affected by other considerations. See the upcoming discussion of vector/sequencing resources, and pattern rate jitter in particular. When in need of narrow and accurate pulses, consider also the effects of driver rise/fall time characteristics and the effects of circuit loading.

Drive Current and Stability: A driver will be capable of sourcing/sinking some amount of current to/from the circuit under test. In the case of in–circuit testing, it may be necessary to overdrive some hefty upstream devices so driver currents in the range of 0.5 ampere may be available. Even this may not be enough. Some devices are difficult to overdrive, but only in one direction. For example, a typical TTL output is harder to force high than low. Here, a solution is to condition such devices, by controlling their inputs, to a state where they are not driving their high–current output. Another approach is to place them into the Hi–Z state if they support it.

Another issue is the stability of the driving signal. Some of the problems here have to do with the problem of backdriving through fixture wiring, which will be covered shortly. Another situation arises when a device being overdriven oscillates. This has been observed with circuits that use TTL levels externally but implement their internal functions with ECL levels. In translating the internal results to TTL output levels, positive feedback is used in the design of the output buffer to enhance translation speed. This can react with the characteristics of a driver (and fixturing) to produce high frequency oscillation.

2.6.2. Signal Quality

The issue of signal quality can very quickly become complex. Factors contributing to this complexity are impedances of sources, impedances of signal paths, impedance mismatches, capacitive loading, nonlinear load impedances, and the like. This discussion will be kept simple. The main point is that signal degradation has the effect of loosening the specifications of your tester's capabilities. Indeed, an order of magnitude change can occur on some specifications. More pathological problems can completely change the nature of your test or cause physical damage to the device you are testing. For the rest of this discussion, assume that we are considering purely digital circuits, and not situations where analog components (especially reactive components) are present on nodes we are testing.

Fixturing: One of the prime contributors to the signal quality problem is the pathway the signals take between the tester and the device under test. The pathway traverses some distance and usually travels through a fixture that provides a simplified mechanical interface to the tester. This pathway should not be thought of as a simple wire, but as a circuit element in its own right that possesses inductance, capacitance, and resistance. In the face of higher frequency signals, fixtures will become a limiting factor in the quest for high quality signal waveforms.

There are three scenarios that influence the design of tester channel resources and the implementation of fixtures. In the first, a driver is applying waveforms to an input of the circuit under test. This input looks like a small lumped capacitance. Good signal quality is maintained by supplying a controlled impedance pathway from driver to circuit input. A 50–ohm impedance is often used.

In the second scenario, the device under test is producing an output to be received by the tester. This output may not have much drive capacity, so it is important to keep the capacitive load presented by the tester and fixture low. This requirement makes a 90–ohm pathway more desirable because of its lower capacitance. Clearly, if a single pathway is used for both driving and receiving, as would be the case for bidirectional pins, a compromise must be made.

In the third scenario, in–circuit testing, a driver is used to provide input to the device under test by overwhelming the upstream driver connected to it. However, this is being done through the fixture pathway. The tester driver must supply a significant current to this upstream driver through the series inductance of the pathway. To really make matters interesting, the upstream driver has a nonlinear source impedance characteristic. This means that as the upstream driver is forced away from its normal level, its effective source impedance will change. The pathway injects delay because of the limit on signal propagation (the speed of light) so that the changing impedance of the upstream driver does not get transmitted to the tester driver instantly. This shows up as strange voltage perturbations on the waveforms seen at the device under test. Normal impedance matching techniques are ineffective in the face of variable terminal impedances. The net effect is that a tester driver with very fast transition time will not be able to create a fast transition *on the board* during in–circuit testing. A 5–nanosecond edge may easily degrade to 40 nanoseconds. Worse, single edges may look like two on the board. (Figure 2.9 shows a signal that is slow rising and nonmonotonic during the time it is rising.) In this case, the nature of the test has changed, destroying the intent of the simulated test.

Clearly, we must take care when preparing a test for an ATE system to consider how closely the waveforms used during simulation will be replicated by the tester. In particular, if we intend to push hard on the limits of the tester and/or the circuit under test, an analysis may be called for or perhaps a physical mockup of the intended fixturing can be studied experimentally. This will ensure that hours of simulation will result in usable tests. See Appendix A for a more detailed discussion of the effects of signal quality.

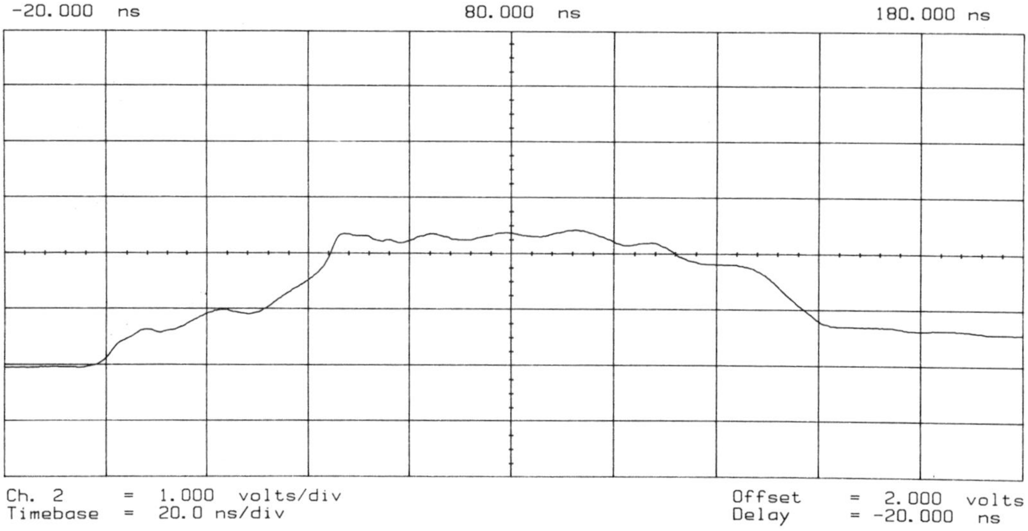

Figure 2.9

Edge Rates: As has just been detailed, stated driver edge rates may not indicate what is truly achievable during testing. One other problem may occur. Again in conjunction with the inductance of fixture pathways, it is possible for high speed signal edges produced by ATE drivers to develop significant overshoot voltages at the input of the device under test, as is shown in Figure 2.10. Some logic families, such as CMOS, will not tolerate having an I/O pin driven to a voltage greater than the power supply voltage or below ground. The silicon structure of a CMOS input under this condition forms a parasitic silicon controlled rectifier (SCR) in the 'on' state, which allows a large amount of power supply current to flow, damaging the device. Voltage overshoots potentially seen during testing could trigger this failure mechanism. Some testers allow the driving edge rate to be slowed down, which minimizes this effect. See Appendix A for a discussion of edge speed and its effects.

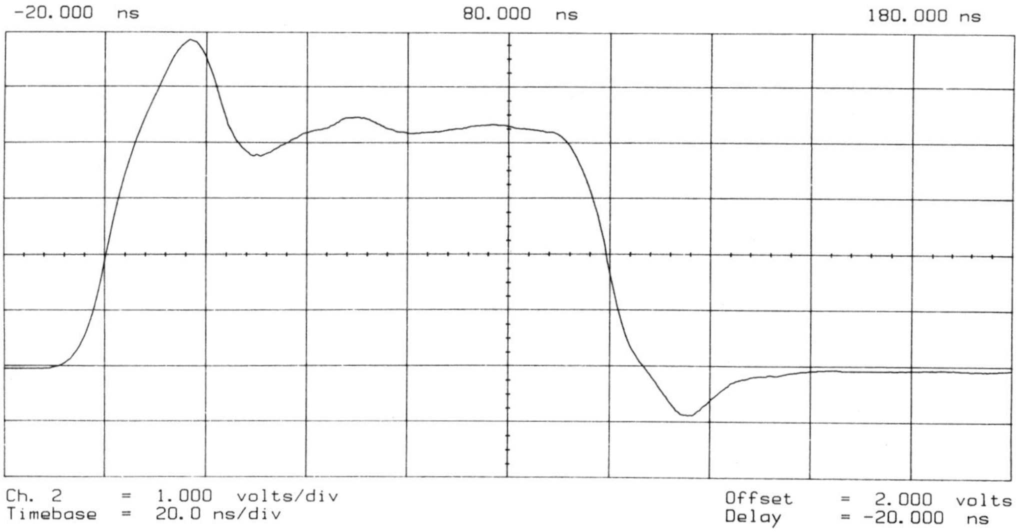

Figure 2.10

Channel Impedance: Channel impedance pertains to the transmission impedance of the pathway from the driver/receiver to the device under test. For high–end IC testers, it is absolutely mandatory to control the impedance of the pathway through fixtures, relay matrices, and the like, because these machines stress timing measurements the hardest. Errors and imperfections in the pathway impedance have the effect of distorting waveforms and loosening the achievable specifications of the tester. Functional board testing targets (as yet) a less aggressive set of timing specifications, but care must still be taken. Again, achievable specifications, not the advertised specifications, must be considered when preparing tests. Finally, during in–circuit testing, the laws of physics take their highest toll on signal quality, as was noted earlier. We should be particularly wary of specifications that are stated 'at the tester' and not at the device under test. Decade degradations on some stated tester specifications must be planned for when preparing tests.

Channel Capacitance: Channel capacitance is primarily an issue when attempting to receive signals from the device under test. This capacitance must be overcome by the device outputs, which may not be very hefty. Some of today's high–end IC testers go to great lengths to reduce this capacitance to less than 30 picofarads. In contrast, expect capacitances of well over 100 picofarads for in–circuit test situations. While functional board test would also have somewhat elevated capacitances, the board drivers being observed are usually designed to drive higher loads. This is not necessarily true when doing in–circuit test, since the engineer has forced his/her way into the middle of a circuit. The effect of channel capacitance is to insert delay into the receive path while the device output charges or discharges the capacitance. This may dictate the upper limit on test cycle speeds.

2.6.3. Vector/Sequencing Resources

In today's testers, the requirement for high vector execution rates, combined with the fact that there are many parallel channels, has dictated that vector information be stored behind the pin in dedicated RAM. Older testers did not have this memory, so each channel had to be updated serially before a new vector cycle was begun. They usually could not execute vectors at rates over a few kiloHertz. Now, there are dynamic devices to be tested that cannot be operated (over full operating range) at rates *below* 8–MHz. This number will probably continue to rise and is a testability issue in itself.

Given that vector information for both driving and receiving is in dedicated RAM behind each channel, the channels are driven in parallel under the direction of a vector sequencer. The simplest sequencer would be a binary counter, broadcasting addresses to the channels to fetch the next vector. However, the nature of tests is that certain vectors are used many times. For example, data may be applied to the circuit under test followed by a standard clock sequence, which is used every time data are applied. This leads immediately to a 'subroutine' concept where data are applied and then a 'clocking subroutine' is called. This allows locations in the vector RAM to be reused, with the net result that much longer tests can be executed than can be linearly stored. This is, after all, the proper use of random access memory. Today's testers have relatively complex sequencers able to generate address sequences for the channels that are by appearances highly random.

It is important to note one complication that may limit the amount of data compression the previous paragraph promises. Let us imagine preparing input for simulation by a CAE system and going to apply a long clock sequence to the device model. It could look something like this in the language of the simulator:

```
for I=1 to  500
     drive 'clock' to 1
     drive 'clock' to 0
next I
```

When this sequence is fed to the simulator, 1000 simulated events then occur. The simulator calculates a result for each of these as it would appear on the device outputs. Now, just looking at the simulated input that represents what a single tester driver must do, there are only two vectors used 500 times each. However, if the tester receiver/comparators are enabled, looking for the circuit responses to each event that the simulator predicted, then there may be many (up to 1000) unique responses. Now, we may require up to 1000 vectors rather than only two, to be stored behind each tester channel because of the unique received information. Is this information desired? Maybe only the last result is important. Only the designers of the test know. They must balance the consumption of vector resources against the utility of each sequence of vectors applied. They may choose not to look at the device responses for some input sequences since they are not as important as others that will come later; these responses will be derived from the *state* the current inputs are setting up in the simulated model.

This brings up an interesting problem: How to instruct the simulator not to display output states that it has calculated but that are considered unimportant? We do not want these unwanted outputs to become the source of unique vectors as the previous scenario presented, because this needlessly consumes vector space. It may turn out that we will have to fix the problem manually. We could do this by editing the output of the simulator or by editing the resultant input to the tester. If our simulator can 'mask' unwanted data, this could be a useful feature.

Maximum Vector Rate: Maximum vector rate is a basic measure of vector sequencing capability. It should not be confused with the frequencies of signals delivered to the device under test because formatting information applied at the channel resources may yield signals higher in *frequency* though not in *information content*. Remembering the previous discussion (section 2.2.1) on formatting, the example of surround–by–complement (SBC) formatting of data yields three edges per vector cycle. A simple unformatted driver can only generate one edge per vector cycle. Thus one vector rate could easily give a three–to–one range on signal frequency. It is more useful to consider vector execution rate as a measure of the information rate delivered to the device under test.

Minimum Vector Rate: A tester may have a minimum rate at which it can apply vectors. This may be dictated by the number of bits within a divider circuit used to synthesize the vector rate from a fundamental oscillator.

Dynamic Range of Vector Rate: Associated with each vector cycle initiated by the sequencer, a vector period may also be dynamically specified. Simpler testers may not allow the period of a vector to be modified on the fly. Fancier testers may allow changes, but from a small, predetermined set of periods. Again, the wide range of periods expected of testers, along with fine resolution, will put practical limits on the *amount* that periods can vary from one vector to the next. This limit is usually several decades wide. It should only be a problem when testing circuits containing one–shots with timeout intervals greater than this range. These will have to be addressed on a case–by–case basis.

Vector Rate Jitter: The term 'jitter' refers to errors in the periodicity of a signal. It can be thought of as a random phase shift of a periodic waveform. The origin of the term comes from the view of a periodic signal on an oscilloscope, where jitter makes the right side of the display appear to quiver. In many testing applications, vectors are applied to the device under test in a periodic fashion and it is generally assumed that jitter in this rate is negligible. This is not necessarily the case.

Vector rate jitter has two differing contributors. One is the basic stability of the hardware used to define or synthesize vector periods. Its immunity to the inevitable noise that exists in the system largely determines how much jitter will appear. On a quality tester, rate jitter should be very low, well under a nanosecond for a period of 100 nanoseconds. Period errors caused by noise induced jitter can be expected to average out to zero over many periods. However, there is another contributor which can inject very significant jitter. It may expand a vector cycle by 50 percent or more and accumulate error over time. This contributor is embedded in the basic architecture of the sequencer.

Some sequencers are implemented as specialized microcode–driven controllers. Their task is to calculate the next vector address for broadcast to the channels. In doing this, they execute a microprogram that may require several microcycles per vector. If this number is always the same there will be no 'sequencer–induced jitter,' and this number will also define the upper limit on sequencing speed. However, in supporting sequencing concepts such as subroutines, looping, and mixtures or nesting thereof, it is possible that some operations, like popping back from a subroutine, will take more microcycles than simply generating a linear sequence of vector addresses. Here, the device under test would show a delay of several microcycles, at, for example, 50 nanoseconds each. This apparent jitter could easily equal the normal periodicity of vector application.

This degree of jitter in vector application is not necessarily a problem. Indeed, many of today's testers inject significant jitter without disrupting the test. One reason for this is that this jitter is much smaller than typical 'keep–alive' times for parts under test. This means that dynamic parts will not forget their state during this unexpected wait interval or 'dead–cycle.' Also, test vectors are typically organized as self–contained experiments, where the drivers cause the occurrence of an event that the receivers look for later. The timing relation from drive to receive is not affected by the injected jitter, or, at worst, the timing is always expanded, never shortened.

However, there are storm clouds looming on the horizon. As an example, requirements are being made that a tester be phase locked to the device under test, typically a board with its own local oscillator. This means that the tester is no longer the definer of periodicity but is required to 'follow' the device under test at its chosen rate. This means that the tolerance for jitter in test periods is greatly reduced. Indeed, the phase–locking process itself can seriously degrade the specifications of the tester's timing capabilities. These factors must be taken into account when planning to utilize external period synchronization and/or low jitter tester cycles.

Vector Synchronization: What is vector synchronization? It is a capability that allows us to hold off an event within a vector until some specified trigger condition occurs. It is used to synchronize a vector with an asynchronous event. Typically, the beginning of the vector period generates an action within the circuit under test which will later cause a trigger condition. This trigger allows the end of the vector to occur, perhaps where a comparison is done. Consider the following simple example: You are functionally testing a dynamic RAM board that has its own refresh control function built in. This refresh controller will occasionally cause a delay in a memory access cycle while it does a refresh operation. The RAM is designed to be accessed asynchronously, using a request/acknowledge handshake.

To test this board, we set up vector definitions that stimulate data, address, and request lines. Then, after the proper wait time that reflects an unhindered access, we set up a wait function that watches for the acknowledge line to activate. Normally, it will activate immediately, and the vector can complete, comparing for the appropriate result. However, if a refresh cycle interrupts the access, our vector will be held up until the acknowledge line finally signals a completion. The net result is that the test sequencer is completely unaware that the interruption occurred, and test vectors can be created as if the RAM were fully static and synchronous.

This raises an interesting question for the user of a simulator who wants to simulate this same board. How do we model this asynchronous refresh function? One way is to assume it doesn't exist or never performs its function. If it were to perform its function at some point during the simulation, then the vectors that result would show the effects of the refresh operation. But, how would you use these same vectors to test the board? You would have to assure that the board and the simulated vectors were 'in phase' at the outset of the test so that when the vectors are expecting the refresh cycle the board actually begins the cycle. If this sounds difficult to actually achieve, it is!

What the CAE user should do with this situation is to assume it does not occur and then use the vector synchronization feature of the tester to hide it when it does. Thus, when we perform our simulation, we 'lie' to the simulator about how this circuit really works. Of course, taking a page from the books of our more scientific brethren, we call the lie a *simplifying assumption*. In essence, we can use a tester feature to get around a knotty simulation problem, without the simulator ever being the wiser. Keep in mind that the performance of the refresh circuitry itself will at some point have to be tested. While not a simulation problem, it remains a requirement for thorough testing.

Vector Depth: When writing a test, we must keep in mind that the tester vector storage resource is finite. It is also likely that we will be testing many different circuits and must archive the data for each, probably on expensive hard disc. While it is true that various techniques such as loops and subroutines can expand the apparent depth of this storage, it is often true that the results of simulation are difficult to map into these structures. Therefore, when using a simulator to prepare tests for ATE, we may find ourselves running short of vector storage space. However, we can compensate for this by using the simulator to produce more efficient vectors (i.e., vectors that focus on real testing issues). This will be addressed in Chapter 5.

Vector/Sequence Compression: Some testers are structured to address the problem just alluded to in the previous section. While the output of a simulator has a highly random appearance to a human, a computer program may find that there are loops and/or subroutine constructs within the data. See [Snook 1984] for a tester architecture that takes advantage of some characteristics of test data to reduce the memory requirements thereof.

It is unfortunate, however, that the amount of compression that we can do with computer analysis or clever architecture is difficult or impossible to predict *a priori*. We generally find out only by trying.

2.6.4. Safety

When testing a circuit, we must take care not to damage it. While this seems an obvious requirement, there are some factors to consider that may impose limits on our testing capability or approach.

First, consider functional testing. Since functional test mimics to some degree the final environment that the circuit under test will encounter, it would appear to be inherently safe. However, not all factors may be similar. For example, some circuits depend on sophisticated cooling techniques that are difficult to implement during testing. In particular, the requirements for guided probing may dictate that the circuit be powered for an extensive period (minutes) while heat removal mechanisms or assemblies are compromised. This may rule out guided probing altogether.

On the other hand, the in–circuit test approach is fraught with risks [Sobotka 1982, Robinson 1983, Bushanam 1984]. As we have seen, poor signal quality can lead to destructive voltage potentials being applied to the circuit. The currents used during overdrive can cause heat buildup within a transistor junction in less than one microsecond [Gravitz 1984, Swent 1986]. The power supply bond wires within ICs can be vaporized by summed backdrive currents as shown in the photograph below. Lifetime degradation [HP 1985] may result when these stresses are less than catastrophic. All of these problems can be successfully overcome [HP 1985], but, in so doing, we may have to impose limits on how we do testing.

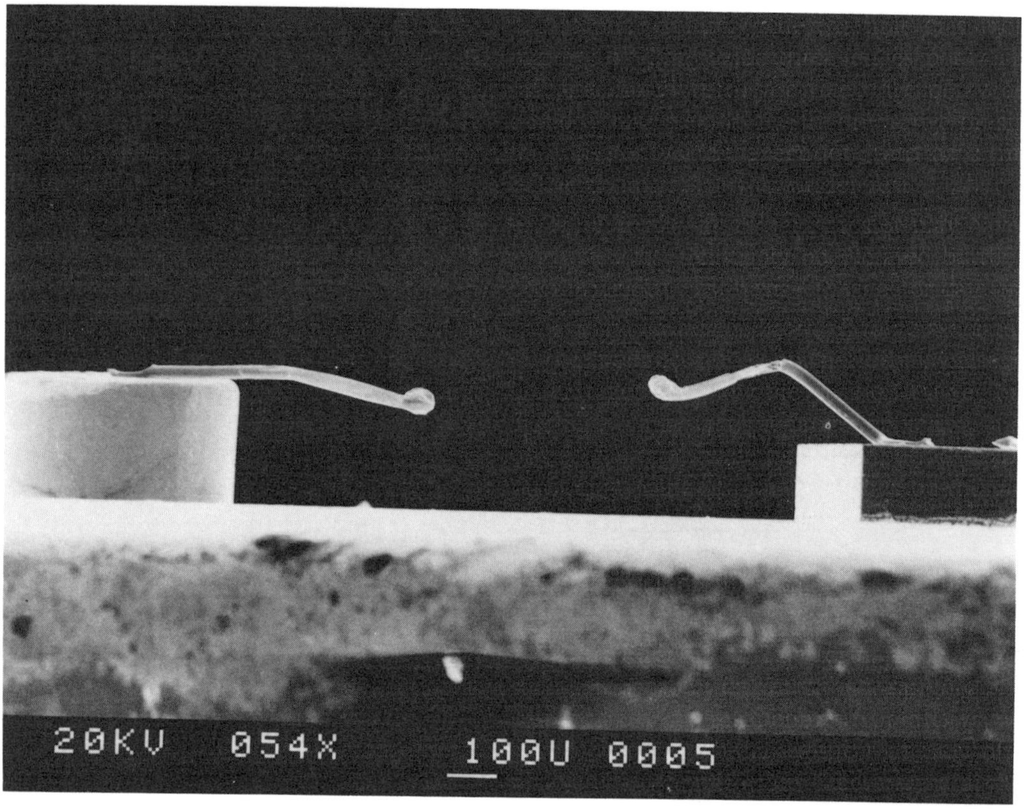

Electron microphotograph of a vaporized IC bond wire. This power supply bond wire has been completely fused by the heat generated by excess current over a short time (1.2 msec). Before vaporization, bond wires may flex and alter position, or they may form cracks and striations. If vaporization does not occur, these other affects may be unnoticed but represent future failure sites. Cumulative degradation caused by subsequent testing may significantly reduce overall lifetime.

2.7. Summary and References

In summary, though digital testers come in various forms and approach the testing problem with different philosophies, they all have the same general architecture. They are a collection of parallel driving and comparing resources. The main differences are largely a matter of specifications with some testers being designed to push harder against the limits imposed by economics and the laws of physics than others. Regardless of whether we are doing IC wafer test or in–circuit board test, the concept of how the tester looks and works is the same.

[Agrawal 1972] V. Agrawal and P. Agrawal, "An Automatic Test Generation System for ILLIAC IV Logic Boards," *IEEE Transactions on Computers*, Vol. C–21, No. 9, pp. 1015–1017, Sept. 1972.

[Bushanam 1984] G.S. Bushanam et al., "Measuring Thermal Rises Due to Digital Device Overdriving," *1984 International Test Conference Proceedings*, Philadelphia Pa., pp. 400–407, Oct. 1984.

[Gravitz 1984] M.E. Gravitz, *The Heating Effects of Backdrive on LS and STTL in the 10ns to 100ms Time Frame*, ATE–Silicon Valley, San Mateo, Calif., April 1984.

[Groves 1979] W.A. Groves, "Rapid Digital Fault Isolation with FASTRACE," *Hewlett–Packard Journal*, pp. 8–13, March 1979.

[Hennie 1968] F.C. Hennie, *Finite–State Models for Logical Machines*, John Wiley and Sons, New York, 1968.

[HP 1985] Hewlett–Packard Co., "SAFEGUARD® In–circuit, a Description of HP's Safe Implementation of Digital In–circuit Test," *Product Note 3065–2*, Sept. 1985, Hewlett–Packard Co., Manufacturing Test Divison, Loveland Colo. 80539.

[Nadig 1977] H.J. Nadig, "Signature Analysis – Concepts, Examples and Guidelines," *Hewlett–Packard Journal*, pp. 15 – 21, May 1977.

[Robinson 1983] R. Robinson, "Investigation of Potential Damage Resulting from Digital In–circuit Testing," *1983 ATE East Conference Proceedings*, p. 25, June 1983.

[Sobotka 1982] L.J. Sobotka, "The Effects of Backdriving Digital Integrated Circuits During In–Circuit Testing," *1982 International Test Conference Proceedings*, Philadelphia Pa., pp. 269–286, Oct. 1982.

[Snook 1984] M.L. Snook and M.A. Teska, "Digital Subsystem for a Board Test System," *Hewlett–Packard Journal*, Vol. 35, No. 10, pp. 14–19, Oct. 1984.

® SAFEGUARD is a U.S. Registered trademark of Hewlett–Packard Company.

[Sohl 1977] W.E. Sohl, "Selecting Test Patterns for 4K RAMs," *IEEE Transactions on Manufacturing Technology*, Vol. MFT–6, No. 3, Sept. 1977.

[Sugamori 1981] S. Sugamori et al., "Analysis and Definition of Overall Timing Accuracy in VLSI Test Systems," *1981 International Test Conference Proceedings*, Philadelphia Pa., pp. 143–153, Nov. 1981.

[Swent 1986] R.L. Swent and M.J. Ward, "Thermal Analysis of Backdriven Output Transistors," *1986 International Test Conference Proceedings*, Washington, D.C., pp. 295–303, Sept. 1986.

[White 1978] E. White, "Signature Analysis, Enhancing the Serviceability of Microprocessor–Based Industrial Products," *Proceedings 4th IECI Annual Conference*, pp. 68–76, March 1978.

[Williams 1983] T.W. Williams and K.P. Parker, "Design for Testability: A Survey," *Proceedings of the IEEE*, Vol. 71, No. 1, pp. 98–112, Jan. 1983.

INTEGRATING DESIGN AND TEST: USING CAE TOOLS FOR ATE PROGRAMMING

Chapter 3
CAE Workstations

Chapter 3

CAE Workstations

CAE workstations are a technology that only recently burst upon the scene. They are largely a product of the price/performance revolution in computing power and the maturing of design automation technology.

The earlier systems were devoted to designers; in particular, designers of ICs. As application specific IC (ASIC) technology has evolved and silicon foundries have emerged, there is increasing emphasis on quick design turnaround. To aid this, CAE tools are extending their domain into the test development area. Meeting the needs of testing was not their original goal, and it will take some time before these systems are more fully tuned to those needs, as well as being readily capable of handling board designs along with ICs.

3.1. Differentiation from CAD Systems

For the purposes of this discussion, computer aided design (CAD) systems are those used for design capture of netlists, schematics, parts lists, physical placements, trace routing, etc.; in short, physical design data. CAE systems are those that utilize logical modeling and allow software–based prototyping to be done, most typically with a logic simulator. CAE systems can provide 'logical' netlists, which describe a circuit in enough detail to perform simulation but usually omit most physical design information. This distinction is historical and essentially artificial but maintained here to avoid confusion. One would expect that future systems would marry the functions of CAE and CAD.

3.2. IC Design

CAE workstations are being applied extensively to IC design in general, and to ASICs in particular. The size of typical ASIC designs is (at this writing) less than 10,000 gates, and most workstations can handle such loads without too much strain. Above this number, simple workstations begin to run out of CPU power. More exotic solutions can be bought utilizing more powerful CPUs and/or hardware accelerators. (Accelerators are hardware coprocessor engines specifically adapted to the needs of simulation.)

As was mentioned before, the primary use of a CAE workstation for IC design is to verify that the intended functionality of the IC has indeed been achieved, both logically and with respect to critical timing parameters. The workstation provides the ability to do 'soft prototyping' and should eliminate a number of design turnarounds. Indeed, reports of ICs working the first time are becoming more common. Faster design cycles are the goal.

However, creating a workable test for the IC is also critical for faster product cycle times. The same model used for designing the IC is an ideal tool for developing an IC test. The verification tests used to prove the correctness of the IC design are a good starting point for creating the IC test. However, verification tests are often focused on problems that are irrelevant at IC test, and also they tend to be incomplete with respect to problems that are crucial to IC test. What often happens is that the original verification test is shortened, relaxed, and then augmented to address those faults or malfunctions that are important at IC test. This process of transforming a verification test into a useful production test is generally difficult for someone unfamiliar with the design. This is one reason that the original designer is more often being required to produce a test.

Board test will be covered in the next section, but let's start with some thoughts: Could I use the model I've developed for IC design in my board design? Could I use the IC test I've developed for in–circuit testing? Are the goals of my IC test the same as the goals of my board test?

3.3. Board Design

CAE workstations are not as yet generally useful for gate level simulation of large, complex digital boards. If a single ASIC can tax a workstation, it is not hard to imagine what a board with 200 sizeable ICs can be like. However, the workstation industry is attempting to overcome present–day limits; this would be a considerable contribution.

Size is but one issue. Another, perhaps worse, problem is that gate–level descriptions for some ICs on a board may not exist, may be incomplete, or may be erroneous. True, you may have models of your custom ASICs done on the same workstation, but the models for the microprocessor, USART, floating point coprocessor, and 802.3 LAN chip are going to be a major headache.

Some CAE workstations have come up with some workarounds for these problems. One is to offer higher levels of modeling abstractions. Regular structures like RAMs or devices with easily described functions like ALUs can often be specified quite easily at a higher level. However, random logic is best modeled at the gate level. An ideal system would allow mixed or 'hierarchical modeling' as appropriate.

Another workaround for modeling ICs that are too large or too undocumented to be successfully modeled is to do 'physical modeling.' Here, a troublesome IC is physically stimulated with inputs that the simulation computes are going to be applied. The IC responds with outputs that are captured by the physical modeling hardware. The captured outputs are sent to the simulator, which uses them as if they were the results of model calculations. This approach is troubled by the fact that the captured outputs are going to be 1s and 0s even though the device may not be producing deterministic results at the sampling point. For example, an output may be in the high impedance state, so that the learned value is just the residual voltage on the nodal capacitance. When doing physical modeling, fault simulation for the logic internal to the device is not possible, although the physical modeler can be engineered to provide I/O faulting capability. Another drawback with physical modeling is that the device may not yet even exist. It would be desirable to develop a model of the device and do simulations in parallel with the production of first prototypes.

Given that a board can be successfully modeled, it is likely, just as in the IC design case, that the intent of the first simulation is to prove that the design is correct. Again, as a soft prototype, a simulation can greatly reduce the design cycle time. Yet, as before, this verification test is created with intents that are much different from those of production test. It is a *strategic* mistake to use a verification test for anything more than the seed from which a good quality production test is ultimately grown.

3.4. Workstation Tools

Most CAE workstations have a collection of tools for use in design, verification, and test developement. Here are some

1. Logical design capture (schematics)
2. Good circuit (unfaulted) simulation
3. Timing analysis and display
4. Race and hazard detection
5. Fault modeling and enumeration
6. Fault simulation and grading
7. Testability analysis and enhancement
8. Automatic test pattern generators
9. Results translators for test equipment

Not all tools may exist on any one station. What follows is a brief description of each, since, though an individual reader may already use a CAE workstation, some of these tools may be absent or may never have been required in the past.

3.4.1. Logical Design Capture

Before a circuit can be simulated or analyzed, a description of the circuit must be entered. This activity is called logical design capture. The design is decomposed into the basic building blocks utilized by the system (a parts list) and the interconnections (the netlist) described. It can be accomplished several ways. On some systems, we do a paperless design by using the CAE system and its graphics capability to directly enter the design drawing. The supporting software generates the resultant parts lists and netlists. Another approach is to download the parts lists and netlist data from a CAD system, where physical layout is being done. Some translation of the data between the two systems is likely to be required.

3.4.2. Good Circuit Simulation

Once the circuit description is entered, and assuming all the component parts have a logical model available, one can begin simulation of the logical behavior of the circuit as it is supposed to behave. This is called 'good circuit' simulation as opposed to fault simulation, which will be discussed shortly.

A good circuit simulator is a powerful tool for design verification. It allows experiments to be run on a design to verify key functionality. It allows variations in the design itself to be quickly analyzed. Since, except for trivially small circuits, design *correctness* cannot be rigorously proven, the design verification process is simply a collection of experiments designed to build confidence that a design meets its goals. For this, simulation is invaluable.

Good circuit simulation can also be used as a 'front end' to fault simulation. Many times, when deriving vectors for fault simulation, we know that a certain state on a key control signal must be achieved before some faults could ever be detected. It may be very tricky to get to this required state that 'unlocks' faults. Good circuit simulation can be used to experiment with vectors until the right sequence is found. Then they can be turned over to the fault simulator.

3.4.3. *Timing Analysis and Display*

To the degree that the CAE system can model the effects of time on the circuit behavior, an analysis of timing behavior can be done with the results displayed in a convenient fashion. Many workstations offer powerful graphical display features and can display timing diagrams as a result of analysis or simulation. There is a wide variation from system to system in how time is treated, particularly since detailed timing studies of large circuits can be costly. Again, for design verification, timing analysis can be an invaluable aid. Since a designer (usually) has a good idea where in a design there is likely to be a timing problem, the scope of study can be narrowed to that area.

3.4.4. *Race and Hazard Detection*

Races and hazards represent serious potential threats to the soundness of a circuit design. As such there is considerable interest in weeding out their effects before committing a design to production. Most systems are capable of simulating the effects of races and hazards. (See [Eichelberger 1965] for a good treatment of this subject.) However, care must be taken on several counts. First, what is the definition of a hazard and how are races identified? The answer differs from system to system based on how they model time. (There will be more discussion of this in Chapter 4.) Second, some systems claim to point out races and hazards for you while others just give results and you must look specifically for them. While the first instance appears desirable, it may actually be of little help if the volume of hazards (in particular) is large and few, if any, present problems. This will often be the case in clocked register transfer style designs containing large combinational logic masses. Here, after each clock cycle occurs, the combination logic may exhibit hazardous behavior. However, in a strictly synchronous design, these cannot effect the final stable state. Having all these hazards displayed would probably represent more of a distraction than a service.

3.4.5. *Fault Modeling and Enumeration*

A major contribution of CAE simulation systems is the ability to analyze circuit behavior in the presence of 'faults.' But, when we acquire a CAE system, we also acquire a philosophy of what faults and a fault spectrum are. Within this philosophy, the system will analyze the model of our circuit and enumerate a list of faults that will be considered on demand. Most systems will allow us to group these faults into classes, such as I/O or internal faults and to consider individual classes only. This is because the number of faults grows very quickly and can easily become overwhelming. Some shortening of the list of faults can be accomplished by removing all but one fault from each class of equivalent faults [McCluskey 1971] (i.e., those with indistinguishable effects). This shortening is useful because it is redundant and wasteful to simulate more than just one fault from each equivalence class. Each will produce identical faulty behavior. However, when later attempting to deliver a diagnosis of a circuit failure, it must be remembered that any fault within a given equivalence class could be the culprit. All should be indicated.

3.4.6. Fault Simulation and Grading

Once a list of faults has been prepared, the effects of these faults can be simulated against a set of input vectors. This is called fault simulation. If the effects of a fault propagate to an observation point (usually a primary output pin) and cause an invalid output (as could be judged by the comparator of an ATE system), the fault is ruled to be detected. Each fault is considered, and the list of those detected or undetected is produced. This 'grades' the effectiveness of the set of vectors in detecting faults.

Most simulators will allow us to continue simulations incrementally to grade the effectiveness of additional vectors against those faults still undetected. This is very useful for two reasons. First, deriving vectors is often difficult. We find that deriving a few and seeing how effective they are is a good approach although it does lead to incremental development of the test. Second, it often turns out that faults not specifically targetted during the derivation of some vectors are indeed detected 'for free.' Hence, if this is known before attention is focused on those faults, considerable savings can result. Fault simulation is costly. It, therefore, pays well to develop strategies to use this resource efficiently.

For general circuit designs, it is not possible to grade vectors in any simpler manner without loss of accuracy. Sometimes this loss is acceptable if a dramatic cost savings comes with it. A nascent methodology called 'statistical fault grading' is just now appearing to fill this need [Seth 1985, Waicukauski 1985], although it is not always available in today's workstations.

3.4.7. Testability Analysis and Enhancement

Beginning in the late 1970s, research in the area of testability measures [Dejka 1977, Goldstein 1979, Kovijanic 1979, Stephenson 1976] began to generate interest. A testability measure defines a scale and a methodology for rating the testability of a circuit. Granting that 'testability' is often a poorly defined concept [Parker 1986], measures are still valuable because they allow us to compare two circuits or two revisions of a circuit. The SCOAP testability measure [Goldstein 1979] was developed in the public domain as a computer program and has proliferated to a number of different computer hosts. Commercial CAE workstations may offer this (or some other) program to allow for the study of the effects of design changes on testability.

One commercially available CAE workstation has recently been advertised in the CAE trade journals that will allow us to enter a design to which it automatically adds testability enhancements. These changes may constitute a fairly radical redesign of the original circuit. After the redesign, it is possible for this system to proceed automatically with test development for the circuit. Until just recently, such capabilities only existed in design systems developed by large, vertically organized corporations. These new systems represent one of the more exciting developments in CAE systems.

3.4.8. Automatic Test Pattern Generators

Automatic test pattern generators (ATPG) utilize the list of faults and the model of the circuit. They analyze the logical and topological nature of the circuit and attempt to create test vectors for each fault. Over 20 years ago, an algorithm was presented [Roth 1966], which will either create a test vector for a fault in a combinational circuit or *prove* that such a vector does not exist. Since then, the search has continued for a more general algorithm that works on any circuit. All that has been generated are heuristic procedures that may produce tests, but may not be successful for all truly detectable faults [Putzolu 1971]. A number of CAE systems offer such heuristic packages. In especially constrained designs such as programmable logic arrays (PLAs) or scan–testable designs [Williams 1983], genuine ATPG capability is possible because the circuits so constrained can be treated as combinational problems. The costs of running any of these programs can be significant [Goel 1980]. The algorithms and approaches employed by these ATPG programs are extensively documented and will not be repeated here. A recent survey of several ATPG packages can be found in [Miczo 1986].

3.4.9. Results Translators for Test Equipment

Currently there is interest in taking simulator results from CAE systems and translating them into a form that could be used for an ATE system. Since the database of the simulator is in some proprietary format and not directly usable on an ATE system, translations are done from the ASCII simulator listing files to a form usable by the ATE system. This is a crude first approach that is highly dependent for success on how well the user of the simulator understands how both the simulator and the ATE systems work. It is also vulnerable to changes in listing formats that CAE vendors change fairly often. Some CAE vendors offer a library of data access routines that we can use to build a translator program. We must still understand how the CAE and ATE systems work.

A few large, vertically integrated companies have had some success in marrying their test development systems to ATE systems from outside vendors. As an example see [Tulloss 1983]. Extensive documentation of AT&T's Diagnostic Organization and Retrieval Algorithms system (DORA) can be found in the conference proceedings from the International Test Conference and the Design Automation Conference starting in 1978.

3.5. Fundamental Limitations

Workstations consist of tools that have fundamental limits. These limits can constrain your choices for testing approaches. For example, you may wish to do functional test of an entire board, but analysis of today's limits on CAE tools may show this to be impractical, either economically or technologically. An understanding of these limitations can be used to make predictions. For example, if we know that our CAE system is hard pressed today to serve our needs, what will a 50 percent increase in circuit density (per year) mean? It is possible that for some tasks a CAE system that is very capable today will quickly become useless after such increases. Our CAE capability is a precious resource to be used wisely.

In this discussion, it is assumed that simple economic limits can be ignored. For example, we will assume that enough RAM exists in the workstation to avoid virtual memory (VM) swapping on disc. VM activity on disc is an implementation detail unrelated to these fundamental limits. Obviously, implementation becomes important when comparing systems and prices.

Limits will be expressed as rising complexity. For example, if a process has a complexity related to the cube of a parameter (stated 'order of N^3' complexity) then we know that as the parameter doubles, the process complexity (read 'runtime') changes by the cube of 2, which is 8.

Good circuit simulation has been observed to have a complexity of order N^1 (linear) to order N^2 (squared). For example, consider a board that is just a series of parallel buffers, such as a bus transmitter. Simulating such a board has a linear complexity as the number of buffers increases. However, a random logic implementation of a sequential state machine tends toward N^2 since added gates add states that may all interact. The interactions are a source of complexity as well as the increase in gates.

Timing analysis is best done on systems specifically architected to model time. Since models are by nature approximations, it is true in some systems that the normal concept of time has been simplified to the point of uselessness. In general, the more accurately the concept of time is modeled, the more complex and expensive the calculations will be. In the most complex simulations done at a 'device physics' level, the calculations involve numerical solutions to nonlinear differential equations.

Race and hazard detection are most accurately done with gate–level simulations and are still subject to basic assumptions about the modeling of time. At higher levels of abstractions, races and hazards are nearly invisible.

The complexity of fault enumeration is a strong function of the fault model chosen. For example, consider faults of the stuck–at 1/0 nature, but not limited to single faults. Then for each signal (or node) in the circuit, there are 3 possible conditions; unfaulted, stuck–0, or stuck–1. Then for N nodes there are 3^N possible fault conditions, a complexity that quickly becomes astronomical. As an example, a 70 node circuit would have 3^{70} or roughly $2.5x10^{33}$ faults. To name each one uniquely would take about 14 bytes each, or just slightly over $3.5x10^{34}$ bytes. The mass of the planet Saturn is about $5.7x10^{29}$ grams, which would be just about enough to store this list at the rate of about 60 kbytes per gram.

But, getting back to earth, limiting faults to single stuck–at 1/0 types makes the number of faults linearly related to the number of nodes, a significant reduction.

Fault simulation has all the complexity of good circuit simulation plus the added dimension that every new gate added to the circuit contributes new faults (here we assume single stuck faults). Therefore, the complexity of fault simulation with respect to gates is between N^2 and N^3. In higher levels of abstraction, it is harder to express the effects of faults in detail and harder to describe how they propagate. However, the level of simplification attained at the higher level may produce good savings. The accuracy of the results is suspect due to these difficulties.

ATPG is, in the general case, still an open problem, as it has been for more than 20 years. For combinational circuits, algorithms exist that will develop test sets, but are time consuming. For sequential circuits, no algorithms exist; there are only heuristic approaches that have limited and unpredictable success and can consume incredible amounts of time even on seemingly simple circuits.

3.6. References

[Dejka 1977] W.J. Dejka, "Measures of Testability in Device and System Design," *Proceedings 20th Midwest Symposium Circuits and Sys*tems, Texas Tech. University, Lubbock, TX, pp. 39–52, Aug. 1977.

[Eichelberger 1965] E.B. Eichelberger, "Hazard Detection in Combinational and Sequential Switching Circuits," *IBM Journal of Research and Development*, March 1965.

[Goel 1980] P. Goel, "Test Generation Cost Analysis and Projections," *17th Design Automation Conference Proceedings*, Minneapolis, Minn., pp. 77–84, 1980.

[Goldstein 1979] L.H. Goldstein, "Controllability/Observability *Analsis* of Digital Circuits," *IEEE Transactions on Circuits and Systems*, Vol. CAS–26 No. 9, pp. 685–693, Sept. 1979.

[Kovijanic 1979] P.G. Kovijanic, "Testability Analysis," *Digest 1979 Test Conference*, Cherry Hill, N.J., pp. 310–316, Oct. 1979.

[McCluskey 1971] E.J. McCluskey and F.W. Clegg, "Fault Equivalence in Combinational Logic Networks," *IEEE Transations on Computers*, Vol. C–20, No. 11, pp. 1286–1293, Nov. 1971.

[Miczo 1986] A. Miczo, *Digital Logic Testing and Simulation*, Harper and Row, New York; 1986.

[Parker 1986] K.P. Parker, "Testability: Barriers to Acceptance," *IEEE Design and Test of Computers*, Vol. 3, No. 5, pp. 11–15, Oct. 1986.

[Putzolu 1971] G.R. Putzolu and J.P. Roth, "A Heuristic Algorithm for Testing of Asynchronous Circuits," *IEEE Transactions on Computers,* Vol. C–20, No. 6, pp. 639–647, June 1971.

[Roth 1966] J.P. Roth, "Diagnosis of Automata Failures: A Calculus and a Method," *IBM Journal of Research and Development*, No. 10, pp. 278–281, Oct. 1966.

[Seth 1985] S.C. Seth, "Predicting Fault Coverage from Probabilistic Testability," *Proceedings 1985 International Test Conference*, Philadelphia, Pa., pp. 803–805, Nov. 1985.

[Stephenson 1976] J.E. Stephenson and J. Grason, "A Testability Measure for Register Transfer Level Digital Circuits," *Proceedings 6th Fault Tolerant Computing Symp.*, Pittsburgh, Pa., pp. 101–107, June 1976.

[Tulloss 1983] R.E. Tulloss, "Automated Board Testing: Coping with Complex Circuits," *IEEE Spectrum*, pp. 38–43, July 1983.

[Waicukauski 1985] J.A. Waicukauski et al., "A Statistical Calculation of Fault Detection Probabilities by Fast Fault Simulation," *Proceedings 1985 International Test Conference*, Philadelphia, Pa., pp. 779–784, Nov. 1985.

[Williams 1983] T.W. Williams and K.P. Parker, "Design for Testability: A Survey," *Proceedings of the IEEE*, Vol. 71, No.1, Jan. 1983.

Chapter 4
Simulators and the Test Problem

Chapter 4

Simulators and the Test Problem

In this chapter, we will cover digital logic simulators which have gate models as their lowest level of detail. While simulators that act at the device physics level do exist, they are not applicable to test development because of the huge sizes of interesting digital circuits and the computational complexity of the physical equations.

Simulation technology has been around and heavily used for well over 20 years. It was first limited to execution on mainframe CPUs by companies with large budgets. See, for example, numerous references to the LAMP system within AT&T that have appeared in test conferences for over a decade. Then an era commenced where ATE manufacturers provided simulators that served as programming front–ends to their digital testers. Examples of these were GenRad's CAPS simulator, Teradyne's LASAR simulator, and Hewlett–Packard's TESTAID simulator. These packages were tightly linked to their respective ATE systems, although LASAR has since evolved into a stand–alone product. They were based on much more modest hardware engines, which brought the cost of simulation down. Now with the advent of personal workstations, simulators are becomming more common as another tool on the circuit designer's bench. Indeed, for rapid, successful development of ASIC devices, simulators are essentially indispensible.

This chapter will help the user of a CAE system judge how to match the features of the user's particular simulator to the testing problem. As will be seen, there are a number of different simulation methodologies. The discussion here is limited to the support of this judgment and is not intended to be a tutorial on simulator technology. There are several texts that cover this technology in detail such as [Chang 1970], [Breuer 1976], and the recent [Miczo 1986].

4.1. Modeling

A simulation is a prediction of how a *model* of a device will behave under user–specified conditions of stimulus. A model is an abstract approximation of the actual device. For example, a gate level model will produce logical 1s and 0s where the real device produces voltages with ringing, overshoot, rise/fall times, current drive, etc. The purpose of the model is to cull out uninteresting physical phenomena and produce usable logical results quickly and inexpensively.

The level of approximation varies from simulator to simulator. Some have no concept of time whatsoever and are only usable for simulations of combinational circuits. Others have higher degrees of timing capability for the price of being more cumbersome. The subtleties of timing can be very perplexing as anyone who has studied (see next page) the 6 Nand gate circuit for the SN74LS74 edge triggered D flip–flop can appreciate.

A simulator can be used to study a circuit without actually having one available. It can also be used to study the behavior of the circuit in one or all of the thousands of possible failure modes. In short, by using a simulator, we can examine a circuit to a detail not practical with a physical prototype.

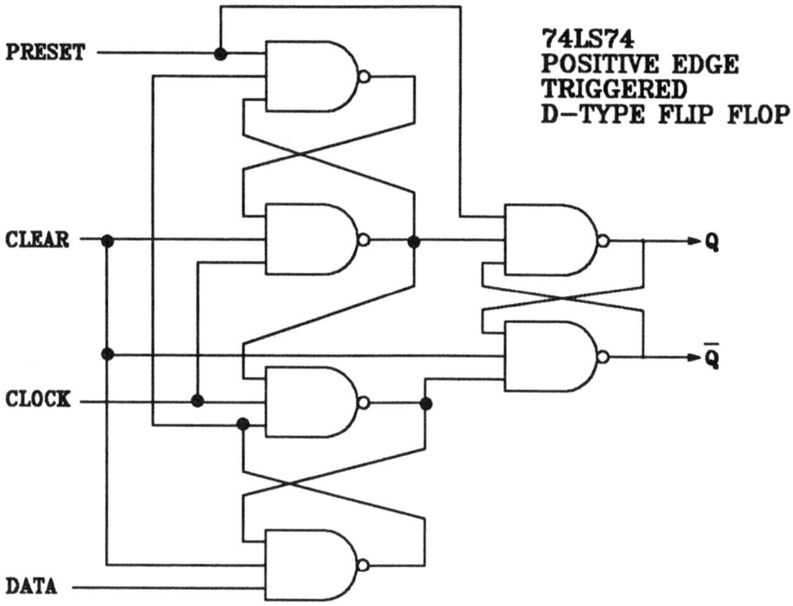

Another fallout of the simulation process is that the data generated are available for analysis after the fact. It is as if one had a physical prototype hooked up to an arbitrarily wide and deep logic analyzer. After running an experiment, one can examine all or parts of the results for hours on end with no need to change experimental setups.

4.1.1. Gate Level Modeling

Modeling at the gate level is done by describing the circuit as a collection of 'primitives' that are wired together. These primitives are (usually) very natural elements such as NAND, NOR, NOT, AND, OR, XOR, etc., which directly mimic the gates the designer is using in the actual design. Other primitives quickly evolved to support other physical structures or concepts, such as CONNECTORs, RAM, ROM, DELAY, etc., particularly since devices such as RAM are tediously impractical to model as collections of gates.

4.1.2. Functional Level Modeling

When the RAM primitive came on the scene, purveyors of simulators were recognizing that simulation at the gate level had its limitations. With the RAM primitive, the two limits addressed are: user patience and simulation run time. A RAM could be modeled with somewhere from five to seven gates per cell, but this comes to over 300,000 gates to represent a single 64K chip. Since a RAM has a very compactly describable function, a 'functional' model was developed that is user friendly, and it is simple to compute its behavior. In some simulation systems, new modeling languages were developed to allow more arbitrary circuits to be functionally described. For example, a multiplier circuit could be thought of as a black box that produces product signals on its output from its input signals by using the well–known rules of arithmetic multiplication.

There are drawbacks to functional level modeling. One is that the behavior of signals inside the black box are no longer visible or available for display. Another is that the level of approximation is much coarser, allowing discrepancies between the model and the actual circuit to creep in. Another problem is that it is more difficult to model and study the behavior of faults (see Section 4.1.7.). Finally, some circuits are just plain hard to model functionally; such circuits are usually collections of 'random logic' that are most tersely described at the gate level.

4.1.3. Physical Modeling

Another approach called physical modeling, has been tried. The types of modeling described so far have relied on software–based models. However, some circuits may contain elements that are very hard to model caused by their size or a lack of documentation. For example, a board may be composed of a hundred 'easy' 74LSxx parts plus a custom LSI processor for which documentation is regrettably unavailable. The board is easy to model except for the processor. A solution to this is to use a special hardware tool that can physically exercise a copy of the processor. Working in conjunction with the simulation software, the 'physical model' calculates with actual hardware the responses of the processor to the stimulus provided by the simulator. These responses are captured by the hardware tool and then 'edited' back into the simulation database. This is a learning process. Iteratively, pattern by pattern, the vectors are increased. Note that the processor has to be driven by all accumulated patterns up to the point where a new one is to be simulated. While the IC in question can be stimulated at a higher rate than is possible in pure software, the iterative learning process can be cumbersome and time consuming.

Certain hazards exist with this approach. First, the IC in question must actually exist and not be subject to revision. Second, garbage may be learned instead of correct data if the IC has not been properly initialized or if it uses buses with periodic high impedance states. Conceivably, the device could be damaged during the learning process if the modeling tool is instructed to drive at an improper time. Obviously, data about internal states are unavailable for display, and fault simulation of the physically modeled device is impossible. Finally, if several ICs on a board require physical modeling, the whole approach soon becomes impractical.

4.1.4. Hierarchical Modeling

Some advanced systems allow more than one modeling technique to be used simultaneously. This allows good flexibility in handling problem cases but can compromise the quality of the result since the limitations of each approach may be compounded. For example, propagating the effects of faults through a gate level description of a circuit is relatively easy. But when these same effects then must be propagated through a functional description of a large RAM, the complexity of the simulation rapidly grows.

4.1.5. Events and Simulation States

Within a logic simulation, signal states must be defined and calculated according to rules. A simple approach is to define two states '0' and '1' and use the well known rules for AND, OR, NOT, etc., to calculate results. However, other states can exist as well. For example, we can represent a high impedance state with 'Z' and the unknown state with 'X.' Now we have to augment our rules (algebra) to handle four states. (See Figure 4.1 for three–valued algebra.) Some simulators have increased the number of states to 9 or 16 to represent other phenomena such as voltage spikes, relative signal strengths, and the like. The cost of this is higher simulation complexity and less ease of use.

Until the advent of the Hi–Z bus, three state (0, 1, X) simulations were quite common. (The addition of the Z state does not seriously complicate matters and will be ignored from here on.) State X can be taken to mean 'unknown' and this is the initial state of every signal until some user–supplied stimulus forces a signal into a known state. But a more important purpose justifies the existence of the X state: the concept of an event.

AND	0	X	1
0	0	0	0
X	0	X	X
1	0	X	1

OR	0	X	1
0	0	X	1
X	X	X	1
1	1	1	1

NOT	
0	1
X	X
1	0

Figure 4.1: 'Truth' tables for AND, OR, and NOT functions in three valued logic.

An event is the changing of a signal value, either directly by user–supplied stimulus or as a result of the propagation of that stimulus. Say a signal goes from 0 to 1. We will assert that this does not happen infinitely quickly, but occurs in two phases: first the signal goes from 0 to X and then from X to 1. A small window of uncertainty exists, although we have not mentioned a concept of time. Thus, we define a signal transition as two events. Now let us examine one consequence of this: we consider a two input AND gate where one signal transitions from 0 to 1 and the other goes from 1 to 0 at the same time. At the start, the output of the AND is 0, as it is when the transitions are done. But we see in Figure 4.2 that during the transitions the output of the gate is X.

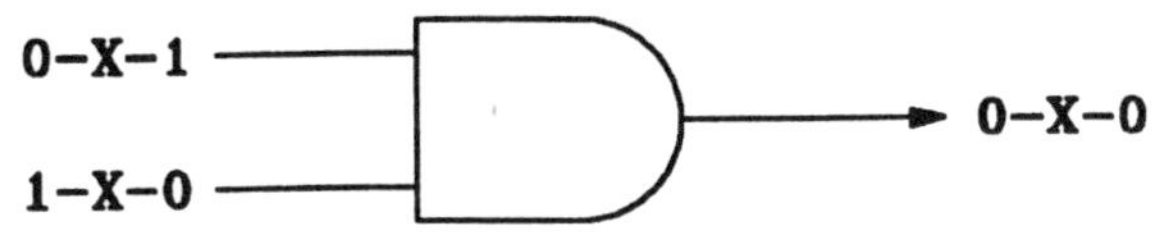

Figure 4.2: An AND gate reacting to multiple input events producing a hazard.

The momentary X is the classical static hazard that such gates are prone to, and yet, we have neither specifically introduced a concept of time in the simulation nor introduced its attendant complexity. It turns out that we can use this event–oriented structure to model sequential circuits, as will be covered later. The point made here is that with the addition of some extra states, more complex circuit behavior can be modeled as a result of some very simple calculations.

4.1.6. Good Circuit Simulation

'Good circuit simulation' is a simulation of the proper (i.e., no–fault) behavior of the circuit in response to stimuli. It is most often felt that fault simulation is the most useful aspect of asimulator. However, good circuit simulation has contributions to make to the test development problem. First is its comparative speed advantage over fault simulation. Analysis of a set of vectors for races, hazards and general usefulness for achieving key circuit states is best done by good circuit simulation. It is also best for preparing 'functional' tests where the faults being addressed are not in the fault simulation repertoire.

4.1.7. Fault Simulation and Test Grading

Fault simulation is a simulation of a circuit in the presence of a malfunction or fault. The fault is chosen with particular characteristics specified by a fault model (see the next section). Usually there are many possible faults. Most simulation systems will automatically enumerate these and, upon demand, inject them one by one into the circuit so that the user can see the effect.

The primary use of fault simulation is to 'grade' a test; that is, produce a score of how many faults are 'detected' by a set of input stimuli (test vectors). A fault is considered detected if the response of the circuit is visibly different from the good circuit with respect to the stimuli applied. Implicit in fault simulation is good circuit simulation with comparison against the faulty circuit behavior. The criteria for 'visibly different' is usually that at least one signal identified as an observation point has reached an incorrect value at some point during the simulation. Differences may occur on other signals, but they may not have been classified as observation points. This maps directly onto testers which cannot observe every internal signal but only those connected to tester receivers.

After a fault simulation is run, one can examine which faults remain undetected and try to figure out new test vectors to append to the accumulated list that may expose them. The process of 'generating' new test vectors is called test generation. To some degree this can be automated, but even relatively simple sequential circuits can be extremely difficult for an algorithm to analyze. Most tests are derived 'manually,' often by trial and error by utilizing a fault simulator for feedback and analysis.

At some point the test is declared complete. This occurs when 100 percent of the faults in the fault list are detected or, more usually, when time, money, and patience have run out. There is the added (frustrating) problem of faults that are logically undetectable. Proving this property of a fault can be very difficult, but until done, one has the nagging feeling that the test is incomplete. Figure 4.3 shows an example of an undetectable fault, an incorrect 0 at point F. The gate with which it is associated is logically redundant but has been added to the circuit to prevent a hazard from occurring.

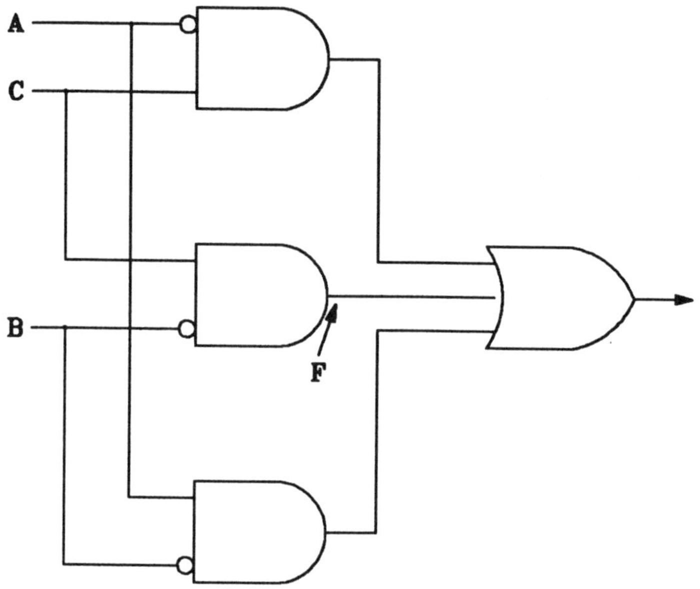

Figure 4.3: A permanent '0' condition caused by a fault at point F in this circuit cannot be detected. The signal is redundantly masked. Whenever it is '1,' one of the other two lines into the OR gate is also '1.' Thus the failing '0' condition cannot pass to the output.

4.1.8. Fault Models

Fault models are approximations of real malfunctions. There are myriads of possible ways that a circuit can malfunction. The purpose of a fault model is to reduce this number to something manageable and relevant. The issue of relevancy is usually determined by experience. For example, if a model is not comprehensive, then a test that is judged to be adequate with respect to this model may prove to be poor with respect to real–world failures seen during testing. The model will have to be altered.

The most popular fault model is called the 'single stuck–at' model. It has been in use for more than 20 years. In this model, a circuit is analyzed for fault sites, which are gate inputs and gate outputs. In general, other points such as positions along a wire are not considered. At any one site, a fault is defined as a gate input/output being stuck at a logical 1 or 0. A circuit of several thousand gates may have 10,000 possible stuck–at faults. During simulation, the effects of only one are active at a time.

Other fault models try to address other possible failure mechanisms. For example, shorts between signals may be approximated by injecting a WIRED–AND function of the signals at the fault site. Failures in memory devices often cannot be approximated by a stuck–at model. Such faults are often attributed to 'neighborhood' effects of other memory cells. Memory devices, caused by their great densities, are essentially impractical to simulate. Fortunately, their regular structure and the nature of their failure mechanisms make it possible to specify tests at a higher, more structural level [Breuer 1976]. This removes the need for fault simulation of memories.

The reason that the single stuck–at model has been so successful is because good test grades found during simulation correlate very well with high yields at the tester. Currently, there is concern that for newer technologies, this correlation will be degraded and a better fault model will have to be found (and engineered into various simulation packages). An example is the stuck–open problem in CMOS technology, where a piece of metalization is missing in the circuit. This can have the effect of converting a combinational circuit into a sequential circuit because of the way that charge can be stored on parasitic capacitances in CMOS. A test set that detects 100 percent of the single stuck–at faults can, with some probability, be applied in a way that fails to expose a stuck–open fault. This problem has generated considerable interest in theoretical circles, but, as yet, no demonstration has been made that shows test yields are suffering. See [Sammuli 1987] for actual yield measurements of CMOS stuck–open faults. In this study, they are infrequent. See [McCluskey 1986] for an excellent discussion of fault models and their problems along with a history of their development.

4.1.9. Limitations

All simulators suffer some limitations when it comes to modeling. Some of the earlier simulators had a difficult time when Hi–Z buses and bidirectional signal propagation became common. Modeling depends heavily on the ability to transcribe a description of a logic device into an interconnected collection of primitives understood by the simulator. For complex components, information may be insufficient for accurate modeling. This makes the task a grueling trial–and–error process.

Simulators often have limitations when modeling and simulating designs containing memory devices. The densities of these devices expressed in 'gate level equivalents' may easily dwarf the sum of all other components. To try to compensate for this, higher–level models or 'partial models' of these devices may be intermixed with the other component models. These work–around approaches may cause subsequent difficulties during test development because the model differs from the actual circuit.

Analog components present yet another difficulty in simulations. Sometimes, problems like these are solved by modeling the effects of the components so that the model is topologically different from the real circuit. For example, an RC delay (Figure 4.4) with two components and three terminals may be modeled as a buffer and labeled with the resistor name.

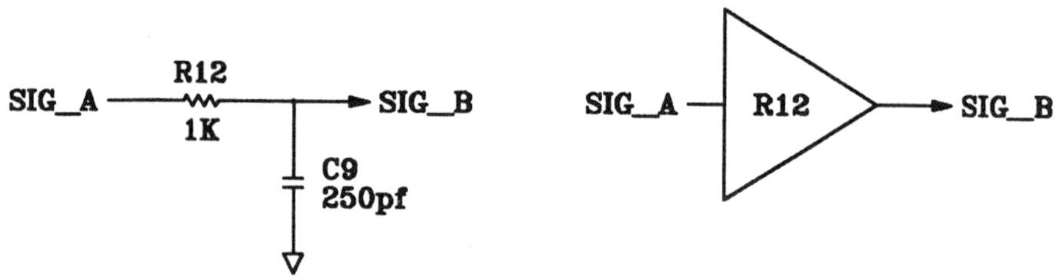

Figure 4.4: Circuit with analog components and one way to model it.

Another choice might have been to model the resistor as a logic delay element. The choice may depend on whether the RC network is being used for a delay that affects the logical behavior of the circuit or whether the RC is being used to smooth glitches on Sig–A. In either case, the model is missing C–9, and R–12 is being modeled with a primitive that has an implicit 'polarity' when the real device is passive. In the end, if a high quality test is desired, C–9 and R–12 should be tested with analog in–circuit tests.

Other common components that cause problems are free–running oscillators, one–shots (particularly those with long and imprecise timeout specifications), relays, RC differentiators, LC frequency doublers, SCRs, Triacs, and transistors. These are very often found in real circuits so plans will have to be made to address them.

4.2. Timing

We have been skirting the issue of timing, one that is quite important in sequential digital circuits. Simulators provide a designer with a tool for analyzing the timing behavior of a circuit. Simulators generally have timing features that address several levels of the timing problem. We have already seen how the choice of state representations and events can be used to model transitional details such as static hazards. The next important level of timing is controlled by the concept of delay. 'Delay' is used to sequence events. Usually, when a pattern is applied to the simulation model, this means that one or more transition events are presented to the simulator. The effects of these are calculated as if they were perfectly simultaneous events (a feat no tester could ever replicate) and a result produced. The introduction of delay allows the user to specify which events are to precede others. This sequencing is critical to getting the simulation of sequential circuits working properly.

Here once again it is stressed that the use of a simulator to study timing relationships is often confused with developing a test for the circuit. A designer wants to *verify* that the *design* is correct. This will involve detailed study of its timing performance. Items like setup/hold times, propagation delays, and other 'nanosecond' concerns are examined. The vectors used to do this in general *are not a good test* with repect to the fault models used for testing, nor are most testers able to re–create the vectors with the same accuracy of the simulation. To summarize, do not use a verification test as a production test.

The following sections cover some of the delay handling schema available in simulators. To use the tester properly and to obtain high test yields, a person developing a test needs to carefully consider the degree that timing needs to be modeled.

4.2.1. Unit Delay

One approach to modeling time is called 'unit delay' simulation. It is used in conjunction with gate-level modeling. Here, each gate reponds to a (sensitized) input transition by producing an output 1 unit of time later. No physical unit (e.g., nanosecond) is attached to this unit. Units simply define the sequencing of events as signals propagate through a circuit. The unit delay concept is somewhat intuitive since real gates introduce propagation delay to signals. Inherent in this approach is a complication having to do with the generation of signal pulses of varying widths.

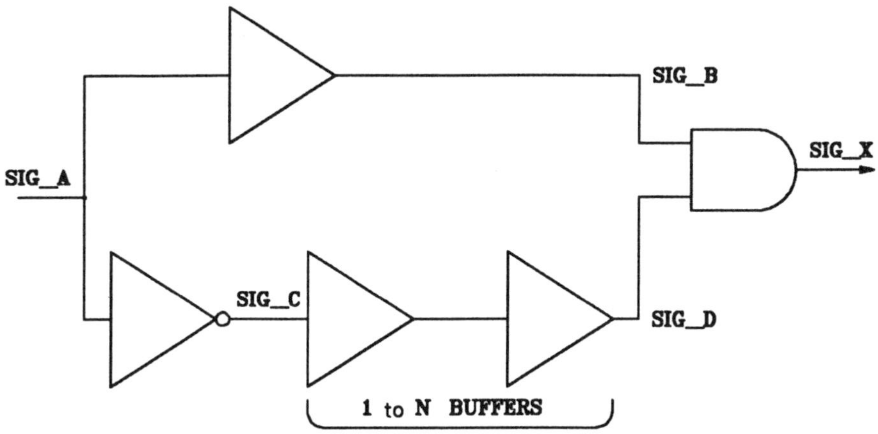

Figure 4.5: Example of reconvergent fanout generating pulses of width N.

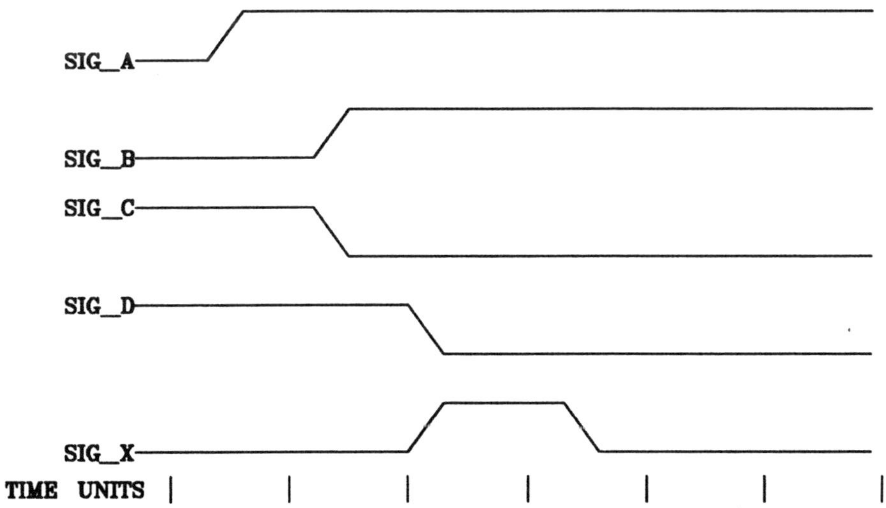

Figure 4.6: Result where N = 1

Consider the example in Figure 4.5 of a signal whose effects fan out through a circuit and later reconverge at a single gate, a very common occurrence. Looking at the signal propagation and timing; the output Sig–X has a pulse that is N units wide, where N is the difference in the number of gates within the two convergent paths. A question then arises on how to interpret the pulse: Is it a hazard, a spike, a narrow pulse, or a one–shot? Could my tester see it? Do I want to test for it? Imagine a case where a signal reconverges after traveling through 20 gates on one path and 22 gates on the other. The unit delay model would predict a two unit wide pulse beginning at unit 22. However, actual circuits would have so much accumulated uncertainty in the two paths that a pulse might not occur at all or, might be much larger than two units. When it started would also be a random variable.

Timing information usable by a tester is not provided (reliably) by the unit delay concept. Delays of importance to a test situation should be modeled explicitly with the use of delay primitives as supplied by the simulation system.

4.2.2. Zero Delay

Zero delay simulations assume that the outputs of a primitive respond instantly to changing inputs; that is, in zero time. There exists a low level delay primitive of one unit (again, not a physical unit) that can be used by the modeler to insert delays in just those places deemed necessary to make the model work logically. It turns out that very few are usually needed. One place they appear is in flip–flops, where they provide hold time between the master and slave sections of the device.

Zero delay models inherently produce fewer pulses during simulation and thus less debate on their meaning. Hazards are shown as 0–X–0 or 1–X–1 events. However, as was the case with unit delay simulation, these units of delay should not be used to simulate meaningful timed events that a tester would want to observe. Use the provided delay primitives designed for macroscopic delay modeling when necessary.

4.2.3. Complex Delay

Some simulators support a delay concept that is quite complex in that a model can be made that very closely represents the actual behavior of a circuit, complete with min/max variations on each timing parameter. There is a run time price to pay, as one would expect. Such simulators often can be instructed to use 'typical' numbers for delay for quicker run times.

With such a simulator, it becomes possible to model circuit timing in a manner that could be useful for testing. For example, with min/max analysis, one could find out where a compare strobe should be set so that all good circuits within these bounds would always test good. One could go even further by modeling the circuit driven by 'tester channels' that have min/max delay times specified that represent tester accuracies and skews. Before going headlong into such simulations, consider how much the additional computation time will be buying in terms of test quality. Perhaps most of the simulation should use typical times with just a few special tests written that really push the limits. What *is* your test strategy?

4.3. Simulation Data

Computers excel at doing mind–numbing drudgery at mind–boggling speeds. Calculating the responses of tens of thousands of gates to tens of thousands of vectors in the presence of tens of thousands of faults fits this description well. If you find yourself complaining about run time, try calculating the amount of data involved and the number of calculations being done.

The simulator generates a lot of data, given the user's description of the circuit (circuit topology), the input vectors, and, optionally, the list of faults to simulate. The results of the simulation are the output states for each input vector, the state of each internal node (those that were modeled), a list of detected faults, and what conditions exist when the fault is detected.

The following sections identify what data is necessary for test, either in–circuit or functional, and what it is used for.

4.3.1. I/O Pin States

The I/O pin states data is a record of the input vectors and the corresponding outputs that were a result. For in–circuit testing, a complete I/O vector consists of circuit inputs followed by a receive delay followed by the expected output. For functional test, there may also be some information about what the formatted channels are doing. The formatted channels utilize timing sets. These conveniently map into 'cycles' (e.g., read cycles, write cycles, etc.). Some simulators support the concept of 'cycles' or 'macros,' which should be relatively simple to map into timing sets if the tester's capabilities are observed.

Note that the I/O pin states are often specified as a linear stream of data for each vector. A position within the stream corresponds to a particular signal, referenced in a signal assignment section. When using a simulator and later translating its output, one must maintain the names of the signals and the correlation of which signal corresponds to which position.

4.3.2. Internal Node States

For functional tests, internal node state information is used for locating faults via backtracing with a guided probe. The data may be learned by the tester in some cases, eliminating the need to extract them from a simulation. There are perils in the learning process having to do with learning garbage states on Hi–Z buses, etc. Since a simulator has the information about the state of every node during the receive phase of a vector, it is a relatively simple matter to collect this information at the same time that the I/O state information is gathered. Note that the names of the nodes will be necessary, and only those nodes that can be probed should be saved. Nodes buried within ICs or deeper levels of simulation model hierarchy are not needed. Internal node state information is not used for in–circuit testing.

4.3.3. Backtracing and Guided Probe Data

Backtracing software is used only during functional testing to isolate a failure with a guided probe. It utilizes internal node state data to determine if a given node has good or faulty signal activity. The other type of information needed is circuit topology. This can be supplied from the simulation database, but one must be careful that the circuit topology of the logical model matches the real circuit topology. As noted in the sections on models, there are opportunities for topology discrepancies. These must be fixed, if they exist, by one of two methods. First, the logical topology can be translated from the CAE system and then hand fixed to match the real topology. A second approach is to declare nodes in the logical model that do not correspond to physical nodes as 'untestable' so that the backtrace package will skip over them and move upstream to nodes that do match the physical topology. This problem can be avoided altogether if the physical design data (from a CAD system) are available, but be careful that you do not get caught with old and outdated information.

An additional and optional body of data that can be derived from fault simulation is fault dictionary information. This information consists of a cross reference of postulated failure mechanisms and the results we would see at the output pins during testing. This is used by an intelligent backtrace algorithm as a 'guess' as to where a fault might be and immediately begins probing at that location. Of course, if the guess is wrong, we can always resume the backtracing algorithm at an output pin that is failing. The guess can easily be wrong if the actual fault that has occurred is not logically equivalent to one of the postulated faults. Be warned that the volume of data that goes into fault dictionaries can be extraordinary.

4.4. Summary and References

This chapter has given a brief overview of simulator technology. There has been much work published in this area. Much of this has had to do with methods of efficiently simulating faults, a subject completely ignored here because it has little to do with ATE, except that inefficient fault simulation may frustrate your desire to grade your tests. The reference by McCluskey discusses fault models and their history. The reference by Sammuli is the first to measure the affects on yield of CMOS stuck–at faults.

[Armstrong 1972] D.B. Armstrong, "A Deductive Method for Simulating Faults in Logic Circuits," *IEEE Transactions on Computers*, Vol. C–22, No. 5, pp. 464–471, May 1972.

[Breuer 1976] M.A. Breuer and A.D. Friedman, *Diagnosis and Reliable Design of Digital Systems*, Computer Science Press, Woodland Hills, Calif., 1976.

[Chang 1970] H.Y. Chang, E.G. Manning, and G. Metze, *Fault Diagnosis of Digital Systems*, Wiley Interscience, New York, 1970

[Chiang 1974] H.Y.P. Chiang et al., "Comparison of Parallel and Deductive Fault Simulation," *IEEE Transactions on Computers*, Vol. C–23, No. 11, pp. 1132–1138, Nov. 1974.

[Eichelberger 1965] E.B. Eichelberger, "Hazard Detection in Combinational and Sequential Switching Circuits," *IBM Journal of Research and Development*, March 1965.

[McCluskey 1986] E.J. McCluskey, "Comparing Causes of IC Failures," *CRC Tech. Report No. 86–2, (CSL TN No. 86–290) Center for Reliable Computing*, Stanford University, Stanford, Calif., May 1986.

[Miczo 1986] A. Miczo, *Digital Logic Testing and Simulation*, Harper and Row, New York, 1986.

[Parker 1979] K.P. Parker, "Software Simulator Speeds Digital Board Test Generation," *Hewlett–Packard Journal*, pp. 13–19, March 1979.

[Sammuli 1987] A. Sammuli, D. Byron, and B. Woodhall, "Empirical Results on CMOS Stuck–open Faults," to appear *1987 International Test Conference*, Washington D.C., Sept. 1987.

[Storey 1977] T.M. Storey and J.W. Barry, "Delay Test Simulation," *14th Design Automation Conference Proceedings*, Computer Society Press, Wash., D.C., pp. 491–494, June 1977.

[Ulrich 1974] E.G. Ulrich and T. Baker, "Concurrent Simulation of Nearly Identical Digital Networks," *IEEE Computer*, Vol. 7, No. 4, pp. 39–44, April 1974.

INTEGRATING DESIGN AND TEST: USING CAE TOOLS FOR ATE PROGRAMMING

Chapter 5
Using a Simulator for Test Development

Chapter 5

Using a Simulator for Test Development

This chapter discusses the use of a simulator for developing a high quality digital test. There are two contexts: in–circuit and functional test. The primary difference between the two is the choice of fault models.

5.1. Strategy

The key to developing a high–quality test is in first analyzing the circuit and then developing a strategy for testing it. Circuits today are intrinsically difficult to test, and getting harder. They contain subcircuits that are difficult test problems by themselves. These problems have to be identified quickly and plans made to address each one. Sometimes more than one approach will exist with some clearly better than others. It is possible that if a tricky testing problem is discovered later in the development, then the work done up to that point will be invalidated.

In addition to the testing problems circuits pose to ATE, they may also present difficulties to simulators. These need to be recognized early and our strategy modified to address them. Our simulator is a precious resource that needs to be used wisely to avoid wasted effort. The following sections cover a list of commonly troublesome problems.

5.1.1. Simulator Capability

Does your simulator system have the capability to get a given job done? 'Capability' is an inclusive term, covering CPU power, memory capacity, disc capacity, addressing capacity, modeling capacity, library support, and hierarchy support. Before beginning a new simulation problem, try to estimate its requirements. From experience, we should be able to extrapolate the new requirements, keeping in mind the effects of complexity discussed in Chapter 3.

5.1.2. Initialization

How is this circuit going to be put into a known state? It may at first seem like a simple question. However, many of today's designs contain RAM, which provides a 'reservoir' of uncertainty (that is, 'X' states) after power–up. It can be tedious in real time to clear it all out, to say nothing of simulating such activity. Yet the presence of uninitialized storage elements can cause nondeterministic test results. To make matters worse, simulators sometimes find that a circuit cannot be initialized. Of course, this isn't really true in the real world (for a valid circuit). It is because of a shortcoming in the way the circuit is being modeled, typically because of the existence of an X but not of a complementary X–bar in the simulator algebra. There is a loss of information at every inverter in the circuit that is propagating an X, since both input and output are X. See, for example, the following circuit in

Figure 5.1 for which the output is known to us but cannot be calculated by a typical three–valued simulator (or other simulator utilizing the concept of an unknown).

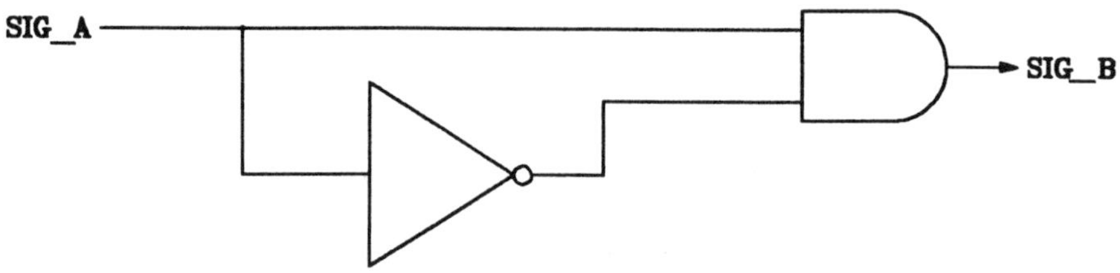

Figure 5.1: Example of circuit with output of 0 for either input.

When Sig–A is either 0 or 1, Sig–B will be 0. But when this uncertainty is modeled as an X, the simple rules of three–valued simulation (seen in Section 4.1.9.) predict the output is also X. We know from the correlation of the two signals presented to the AND gate that a 0 must occur. However, when less trivial circuits exhibit reconvergence with multiple sources of uncertainty, it becomes impractical to calculate the result. (Each new X in the circuit doubles the number of potential interactions that must be analyzed.) Because of this 'lossy' nature of a simulator system, a model may be impossible to drive into a known state. The only solution is to *assume* an initial state. Then, for this to be a valid seed for test development, the real circuit must start from the same state during a test.

Here we must discuss a fine point: It is not strictly necessary to have all storage elements at a known state at the beginning of a test. A core of key storage elements will suffice. For example, in a microprocessor board with RAM (for program store), the microprocessor's instruction decoding state machine must be initialized. Then the microprocessor can fetch instructions from its boot ROM, which allows the microprocessor to initialize a section of RAM, perhaps with a test program. This part of RAM can then be fetched as the test program is executed. Since we have not put the entire RAM into a known state, it is technically incorrect to call the circuit 'initialized,' and, indeed, until it is, a completely thorough test is impossible. However, especially with respect to costly simulation, practical limits will usually prohibit full initialization.

If the circuit to be tested cannot be readily initialized and particularly if initial state assumptions must be made, then the initialization strategy must be carefully thought out. Some tests that are branded as 'flaky' or circuits that are described as 'intermittent' are really exhibiting the effects of incomplete or inaccurate initialization.

When it is necessary to assume an initial state from which to begin simulation and you cannot alter the design to fix the problem, it will be necessary to add to the resulting test a homing sequence as a preamble to the test. This homing sequence will have as its goal the production of the same state the simulation assumed among the core storage elements. Writing a homing sequence can be a difficult task since your CAE tools are of no help at this point.

5.1.3. The Effects of Faults on Initialization

It is often a nontrivial problem to initialize a circuit. A class of design for testability techniques exists to remedy this [Williams 1983]. However, a circuit that is easily initializable *when operating correctly* may still exhibit seemingly unusual behavior in the presence of faults. Consider the simple circuit in Figure 5.2, consisting of a 2–bit counter with clear and parallel load capability and an AND function that indicates that the counter has reached '11.'

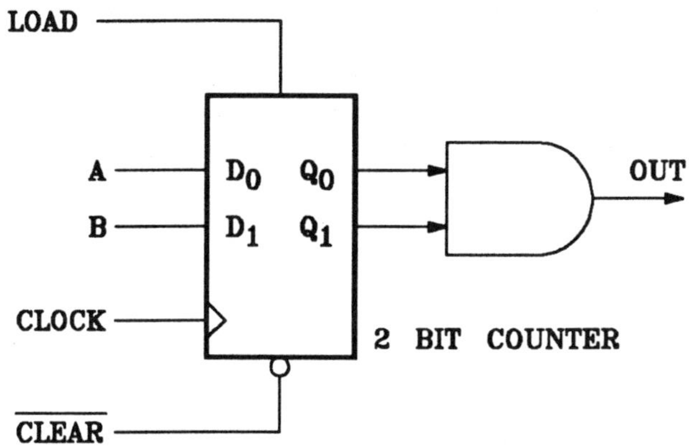

Figure 5.2: This simple and highly testable circuit may exhibit seemingly intermittent behavior in the presence of a hard fault, with respect to the test in Table 5.1.

We have free access to all the input lines so this circuit passes as highly testable. Table 5.1 shows (in a functional form) a test and the resulting output of the good circuit, as well as the output of the circuit with a fault during two runs of the same test. An asterisk marks each faulty output.

Table 5.1

Action	Good Output	First Bad	Second Bad
Clear	0	0	0
Count	0	1 *	0
Count	0	0	0
Count	0	0	1 *
Count	1	0 *	0 *
Load 01	0	0	0
Load 11	1	1	1
Load 10	0	0	0

In this example, there is a single fault for both fault behaviors, an open trace (broken wire) between the CLEAR signal and the counter. Assume the counter input floats high, looking like a stuck–at 1 that disables the CLEAR function. Insidiously, the *good output* can also occur in the presence of this fault. The key to this behavior is the initial state of the counter. If the counter is in state 00 when the test begins, then the fault will not appear on the output because the first action is to clear the counter to 00. Thereafter, the circuit behaves normally. If the test is rerun, hence starting from the state left over from the first run (10), then the first bad output is observed. The second example of bad output would occur if the counter started in state 01. Indeed, there are four possible states from which to start the test. In the presence of the CLEAR fault, there would be three different ways the test could fail and one way that it would pass! Now let us try a different test, a simple reorganization of the test above, shown in Table 5.2.

Table 5.2

Action	Good Output	CLEAR Stuck 1
Load 11	1	1
Clear	0	1 *
Count	0	0
Count	0	0
Count	0	0
Count	1	0 *
Load 01	0	0
Load 10	0	0

Now, the circuit always exhibits the same behavior in the presence of the CLEAR stuck–at 1 fault. The initial state no longer matters. The reader is encouraged to examine the behavior of the same two tests for the case where the LOAD signal is stuck–at 0 (i.e., is inoperative). In either case, does the initial state produce a set of possible responses? Can a test pass in the presence of this fault?

Notice that the two tests both assert CLEAR once and LOAD three times. Each function is capable of initializing the properly operating circuit. The difference is that the first test uses a single CLEAR as the initialization, while the second uses the multiple occurring LOAD function. Further, the two alternative forms of initialization are used at the beginning of the test and produce initial states that are *visibly* different from the view of the output signal. Even this may not be enough: What if the first form is inoperative due to a fault but is not reissued later in a way that visibly alters the state of the circuit? The surest way of initializing this circuit would be to issue first one form, then the second, and then the first again: for example, CLEAR, LOAD 11, CLEAR. This will assure that faults that disable one or the other initialization cannot be masked by a randomly occurring starting state.

Some simulators offer a special feature when performing a fault simulation. They note when a fault causes a circuit output to be unknown (X) when, in addition, the good circuit is known, either 0 or 1. This condition is termed an 'X–detect' or 'possible detect.' This feature can be used as a warning that the above problems of seeming intermittency or nondeterminism can occur on your tester for the test as written. If so, it may be wise to rewrite or augment the test to minimize these warnings. A discussion of how a simulator can manage X–detects may be found in [Parker 1979].

5.1.4. Asynchronous Interactions

Many circuits designed in the past have been 'clocked synchronous' designs. Today, particularly at the board level, it is possible to place several major system components into one board that will require testing. Each major subsystem may have its own clock distribution system. The subsystems communicate among themselves with asynchronous handshaking schemes. For example, (see Figure 5.3.) consider a microprocessor (with its clock) that talks via an asynchronous handshake (like DTACK) with a memory subsystem that has a memory management unit (with its own clock of different frequency). This MMU will occasionally hold off DTACK while doing refreshes or arbitrating memory accesses. What should our strategy be for using a logic simulator to develop a test for this circuit?

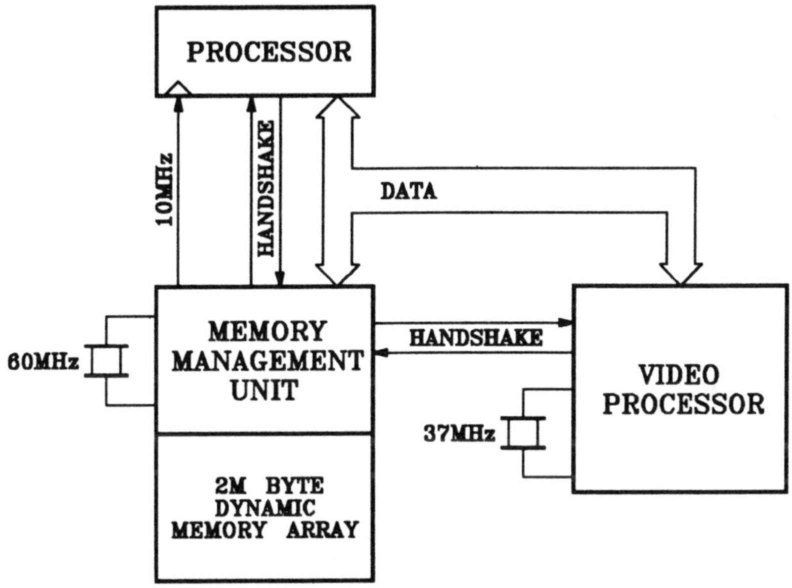

Figure 5.3: This simplified drawing shows a circuit with examples of two types of asynchonous interaction. The MMU/memory array runs from a 60–MHz clock. Internal refresh counters embedded in the MMU will occasionally impede memory accesses by the processor, although it derives its 10–MHz clock from the MMU. The video processor runs from its own 37–MHz clock, which has no phase relationship with the MMU clock. The three subsystems interact with two-wire handshake protocols. This example is taken from a 32–bit workstation design with the core system hardware contained on one printed circuit board.

It is very hard to model the actual behavior of this circuit because there is no fixed phase relationship between the two clocks. One approach is to make a simplifying assumption that an asynchronous handshake will take a precise amount of time to return. That is, the response will always take *n* clock cycles. We then simulate this behavior, and the test development engineer manages the consequences. It is possible in some situations for the test engineer to mimic this behavior during the test by using a tester feature such as vector synchronization, discussed in Chapter 2. He/she would set up the tester to look for some event to occur (like DTACK) during each test cycle. If DTACK is late, freeze the current test cycle while letting additional clocks happen. When DTACK does occur, continue the test.

This approach can work *if* those additional clock cycles do not perturb a (visible) state of the circuit such that the real circuit and the simulation are now out of synchronization. For example, the additional clock may go to the MMU where a dynamic refresh is under way. When complete, the state of the memory subsystem is the same (again) as the simulation. However, the clocks may also have gone to an interrupt timer that progresses slightly closer to its timeout limit. Since the internal state of this timer is not visible to the tester, we can continue, but eventually it will time out sooner than the simulation predicted. This could look like a fault unless we now account for this possibility. Other internal discrepancies may occur. The risk is whether we have thought of them all.

Another strategic approach to this problem is with a divide–and–conquer method: Treat each asynchronous subsystem as a functional cluster and test it separately. This allows extensive testing of the workings of the cluster without the synchronization problem of the real world with the simulation. We should be wary, though, when testing one subsection that spurious handshakes do not return from adjacent subsystems and 'sneak' past the overdrive levels in the form of short pulses. (See Section 2.4.2.) Then we create an additional test that exercises the circuit as a whole to prove the asynchronous subsystems successfully communicate. This last test may not even require simulation but may easily be written manually.

5.1.5. Unmodeled Circuitry

Is there circuitry that cannot be practically modeled? Imagine, for example, a circuit which has an analog–to–digital converter (ADC). We could provide an analog signal as an input to the ADC, but we run the risk of indeterminate behavior on some (or even all) its digital outputs if the analog signal is close to an LSB (least significant bit) boundary. Consider this 12 bit case expressed in octal: say the analog signal given to the ADC produces 3777. With another ADC the result might be 4000 because of a slight difference in the way the LSB looks to the two ADCs. The circuit is functional to within a half–LSB and yet 12 digital signals are 'wrong.'

Before going to the trouble of modeling and simulating a circuit, we should carefully examine the parts list and topology for things that are going to be difficult. We should decide on a test strategy that takes these problems into account. Often, we will either excise the offending circuitry with in–circuit overdrive or create a fictional model that departs from the real circuit but approximates the overall behavior of the problem circuitry. This approximation must be treated like a black box and should be carefully documented as to its purpose and limitations.

5.2. Fault models

One of the main contributions of a simulator is the ability to do a fault simulation and to 'grade' the effectiveness of a proposed test against a list of faults. These faults fit into a model of how the circuit is expected to fail and are in themselves an approximation of real–world physical failures. The only true validation of a fault model comes from an analysis of the yield of the testing process. How many real–world failures escape detection by the test prepared with respect, to a fault model? Another question seldom asked but that is related is: how many faults must one model to achieve good coverage of the real–world failure mechanisms? This is similar to asking, are we testing too much? This normally is not an issue unless the amount of overtesting being done is lowering total test throughput, consuming excessive (disc) storage space, or consuming inordinate simulation resources.

5.2.1. In–Circuit Test

In–circuit testing inherently offers very high levels of diagnostic resolution. The fault model used is the 'manufacturing' fault model. This model assumes that one begins with perfect components and builds a board with an imperfect process. Various faults will occur because of these imperfections. The physical mechanisms that frequently occur are

1. Bent or broken IC pins, lack of solder (no contact or open)
2. Solder bridges (shorts)
3. Improper orientation (component or polarity reversed)
4. Wrong component inserted
5. Electrostatic discharge damage
6. Missing component

The fault model used (typically) to model the majority of these failure mechanisms is the 'I/O pin fault.' Here, the I/O pins of each IC are considered to be potentially stuck–at 1 or 0. No specific mention is made of the faults that could exist deep inside an IC. To detect the output pin faults, one merely makes each output achieve both 1 and 0 states. To detect all input pin faults is more difficult. One has to set up tests that will prove at the outputs that each input is free to achieve both states. Simulators are very useful for this task. One merely specifies that the faults to be simulated are the I/O stuck–at faults (most simulators break the faults down into classes such as these). The list of faults is only two times the I/O pin count, which is small and not growing anywhere nearly as fast as the numbers of gates being integrated inside the IC.

What about coverage of the manufacturing faults? First we recognize that solder bridges are eliminated by a nondigital 'shorts' test done before digital testing is initiated. Shorts testing is done without applying power to the board and thus potentially hazardous shorts will be eliminated before power is applied. Some of the other mechanisms, such as improperly oriented parts, produce such gross malfunctions that even a most cursory test will observe them. A few are more subtle, like open pins and electrostatic discharge damage, but these are specifically addressed by the fault model.

Perhaps the most subtle problem to find is the wrong component problem, which is found in rare cases. Surely, if a dual flip–flop is replaced by a quad–OR package, almost any test will detect it. But if a quad–OR test designed to catch the I/O pin faults is presented with a quad–XOR function, there is a chance that the test will pass. Sometimes, the specific I/O pin fault model is augmented with additional tests that will detect such problems. These are usually added to the original test when test yield analysis proves that a failure mechanism is incompletely covered.

5.2.2. *Functional test*

Functional testing (of boards or clusters) does not offer the same degree of diagnostic resolution as in–circuit test. However, because of fixturing difficulties, it may be the only type of test possible. A simulator can be a great deal of help because each functional test must be custom generated.

The fault model most often used is the single stuck–at model. Practical limitations will only consider those faults at the 'main level,' which means those faults occurring at inputs/outputs of all ICs. Even with this limitation, the number of faults under consideration can number in the many thousands for typical boards.

It can be a strenuous task to excite a single stuck–at fault and propagate its effects through a maze of clocked logic to a point where it can be observed. In so doing, a side effect can be that other faults are carried along for the ride and detected 'for free.' It should come as little surprise then that faults not specifically modeled, such as pin–to–pin shorts, are often detected for free. While it is known that the single stuck–at model is theoretically inadequate [Mei 1974] as a test for shorts, experience with test yields has for years shown that shorts are usually covered quite well. Simple tests for shorts are sometimes added where physical circuit structure suggests that they could be a problem. For example, buses are often routed across boards as multilane highways. Shorts across bus wires are a potential that can be discovered by sending a few patterns of alternating 1s and 0s down the bus. Another class of 'unmodeled' fault are those deep inside the ICs. If one believes the ICs were perfect before board fabrication, then the fault spectrum in question is virtually the same as that for in–circuit test. Again, the amount of exercise that a board gets during a test for all stuck–at faults seems to detect a large number of faults that do not match the model. Test yield analysis, again, gives the true measure of test effectiveness and will point out when assumptions about fault models need improvements.

5.3. Observing Tester Limits

A simulator, being a software environment, is not limited to the laws of physics or the laws of economics, at least to the extent that a tester is. Therefore, it is quite easy to do something to a simulated circuit that cannot be reproduced by the tester. Perhaps it will not matter, but more often than not, the tester will not succeed in producing the same environment and the result will be a failed test.

The best way to stay within the limits of the tester is to plan for those limits during the development of the test strategy. It is generally too late when simulation data are created without constraint; it is impractical to try to modify the data (most likely by hand) so that the results on the tester again match up with the patched up simulated predictions.

If a simulation session is set up within tester limitations for circuit inputs, and the minimum cycle times and resolutions are pressed hard on circuit inputs, it is possible that the circuit responses will not map into the response monitoring capability of the tester. Ideally, the rules and realities imposed by the tester should be used as guidelines for setting up circuit stimuli. Then the results of the simulation should be checked to see that the simulated circuit responses can be mapped into tester capabilties.

Since simulation can be time consuming and expensive, it is advisable to run some quick experiments with the good circuit simulator that prove that the specified circuit stimuli and responses stay within tester limits. If these experiments are used as templates for building and assembling tests, then a great deal of wasted effort may be avoided when test generation and fault simulation are done.

5.4. Test Generation Methodology

For meaningful circuits (that is, large and sequential) there is no automated way to generate tests in the general case. Some software packages will attempt to develop tests for specific faults by using a set of heuristic procedures [Lit 1986]. Their success fluctuates, and failure to find a test does not imply that one does not exist. Often, when a test is developed, it is simulated against the set of faults that remains at that point to see if other faults are detected 'for free.' For combinational circuits or those that can be treated as combinational (such as those designed with the LSSD discipline) [Williams 1983] algorithms do exist that will either find a test or prove that a test does not exist for a given fault. (See the well known D–algorithm [Roth 1966, Roth 1967].) Again, once one test is found, it is simulated against all faults to see if others are also detected. This is because the cost of the analysis for test generation is greater, in general, than the cost of fault simulation.

Most tests are created 'manually' by the simulator operator. A given fault is targetted and a sequence of inputs is postulated that will excite the fault and drive its effects toward an output where it can be observed. Then the simulator is called to see if this has actually been achieved. Other faults may be detected by the same sequence 'for free.' A simulator gives feedback about internal node states that can be quite valuable when trying to force a stubborn fault into a detected state. One strategy for generating tests is to attack 'stubborn' faults first, although the ability to categorize faults comes with experience, or from testability measurement algorithms (see [Goldstein 1979] and [Kovijanic 1979]). In general, faults located far from inputs and outputs and that require very specific memory states to 'unlock,' are stubborn. When a test for such a fault is found, it is likely that others nearby will also be found during simulation.

Recognizing that the analysis of faults for test generation is very complex, there has been research into alternate approaches to the problem. For example, at the University of Illinois work has been done on test generation for microprocessors [Thatte 1979]. This work is extendable to general designs if they are easily described with register transfer models. A distinct advantage of this approach is that it does not rely on the existence of detailed gate–level design information. In real–world testing, this information is often missing. Another approach is random test generation. Here, the cost of producing vectors is very low, but there is no indication of their effectiveness without some form of grading. A simulation may be used to see what faults are detected, just as is often done with deterministic test generation. In this case, we may come up empty and have to try other vectors. See [Agrawal 1972] and [Shedletsky 1975] for reports on the tradeoffs. See [Williams 1977] for an example of how random testing may be combined with LSSD discipline. See [Parker 1976] for a methodology for enhancing random test generation efficiency. See [Losq 1976] for a novel approach to the design of test equipment by using random test.

A balance may need to be struck between short test sequences that detect many faults quickly and longer test sequences that have lower detection rates. The tradeoff is driven by the ease of fault isolation. A short test costs less to develop (in terms of simulation time) and will run somewhat more quickly on the test hardware or, more important, not put as much stress on the tester memory resources. However, when isolating the exact cause of a problem, if each input vector may flush dozens of possible faults, then the resolution per vector is low and additional measurements (like probing) will be needed. For in-circuit testing, diagnostic resolution is implicit, so a short test is best. For functional testing, longer sequences will give better resolution with faster diagnosis, unless simple go/nogo testing is the only requirement.

5.5. References

[Agrawal 1972] V. Agrawal and P. Agrawal, "An Automatic Test Generation System for ILLIAC IV Logic Boards," *IEEE Transactions on Computers*, Vol. C–21, No. 9, pp. 1015–1017, Sept. 1972.

[Goldstein 1979] L.H. Goldstein, "Controllability/Observability Analysis of Digital Circuits," *IEEE Transactions Circuits Systems*, Vol. CAS–26, No. 9, pp. 685–693, Sept. 1979.

[Kovijanic 1979] P.G. Kovijanic, "Testability Analysis," *1979 International Test Conference*, Cherry Hill, N. J., pp. 310–316, Oct. 1979.

[Lit 1986] The Literature 1986. See various advertisements in the following trade publications in 1986

IEEE Design & Test of Computers (Computer Society of the IEEE)
Electronic Design (A Hayden Publicaton)
Electronics Test (A Miller Freeman Publication)
VLSI Systems Design (A CMP Publication)

[Losq 1976] J. Losq, "Referenceless Random Testing," *1976 Fault Tolerant Computing Symposium Digest*, Pittsburgh, Pa., pp. 81– 86, June 1976.

[Mei 1974] K.C.Y. Mei, "Bridging and Stuck–at Faults," *IEEE Transactions on Computers*, Vol. C–23, No. 7, pp. 720–727, July 1974

[Parker 1976] K.P. Parker, "Probabilistic Test Generation," Ph.D. Thesis, Electrical Engineering Dept. Stanford University, Stanford, Calif., June 1976.

[Parker 1979] K.P. Parker, "Software Simulator Speeds Digital Board Test Generation," *Hewlett–Packard Journal*, pp. 13–19, March 1979.

[Roth 1966] J.P. Roth, "Diagnosis of Automata Failures: A Calculus and a Method," *IBM Journal of Research and Development,* No. 10, pp. 278–281, Oct. 1966.

[Roth 1967] J.P. Roth et al., "Programmed Algorithms to Compute Tests to Detect and Distinguish between Failures in Logic Circuits," *IEEE Transactions on Electronic Computers,* Vol. EC–16, No. 5, pp. 567–580, Oct. 1967.

[Shedletsky 1975] J.J. Shedletsky, "A Rationale for the Random Testing of Combinational Digital Circuits," *Proceedings of COMPCON*, pp. 5–9, Washington, D.C., Sept. 1975.

[Thatte 1979] S.M. Thatte, "Test Generation for Microprocessors" Ph.D. Thesis, University of Illinois, Urbana, Ill., May 1979.

[Williams 1977] T.W. Williams and E.B. Eichelberger, "Random Patterns within a Structured Sequential Logic Design," *1977 Semiconductor Test Symposium*, Cherry Hill, N.J., pp. 19–27, Oct. 1977.

[Williams 1983] T.W. Williams and K.P. Parker, "Design for Testability: A Survey," *Proceedings of the IEEE*, Vol. 71, No. 1, pp. 98–112, Jan. 1983.

Chapter 6
Translation Issues

Chapter 6

Translation Issues

It is often the case that the simulator used to prepare the test is not explicitly designed to work with your ATE system. Indeed, it is not unusual to see two (or more) different simulators used to develop a test for a complete board. The in–circuit vectors for some gate array may come from a workstation, while a large cluster of chips may be simulated on a large capacity mainframe–based simulator. In any event, the result of the simulations must be postprocessed into a form the ATE system will accept.

Assuming the simulations have been done with the ATE system capabilities in mind, translated vectors should work. However, there may be some actions required of the user before they really do. These will be covered in this chapter. If violations of ATE capability have been committed, then the translated results are not going to work. Sometimes the translation process itself will detect potential problems. Sometimes, long and frustrating debug sessions on our expensive ATE system are the only way to find these errors.

A summary of the issues covered here and throughout this book are compiled in the form of checklists in Appendix B. These checklists should help us ask the necessary questions to avoid the many problems that may occur.

6.1. Types of Translation Approaches

There are a number of approaches to the translation task. The one chosen is often dictated by the level of support that the simulator vendor is willing to provide.

In the worst case, no support is offered. Here, let us assume that the host computer that runs the simulation has programming support (i.e., a compiler, editor, linker, library support, etc.). Then we can attempt to build a simple reformatting translator that uses the output listing files of the simulator as its source, as shown in Figure 6.1. This effort may be complicated by certain facts of life. First, simulator output reports may have variable formats. Indeed, some systems let users personalize the output listings. Also, simulator vendors sometimes change output formats during software updates. This means we have to deal with a moving target. Second, the very nature of 'listings' is that they are limited to a relatively small number of columns of information, usually between 80 and 120. Since many circuits of interest may have more than 120 signals and/or 120 vectors in the test, this means that data represented on the listing are incomplete and several listings may be needed to display all that is going on in the circuit. Thus a translator may have to gather and organize data from several sources.

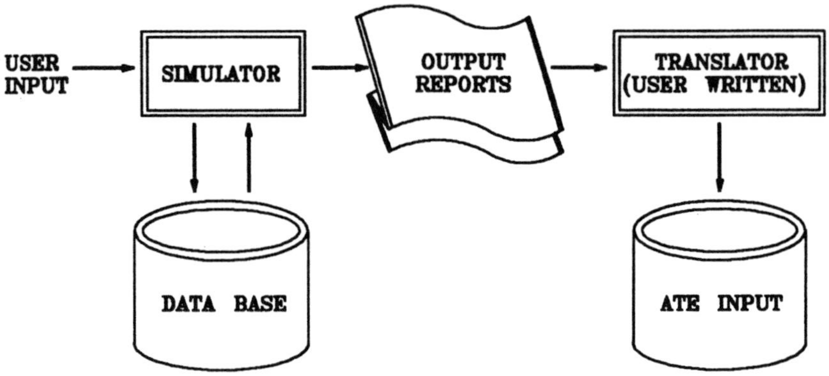

Figure 6.1

In a better case shown in Figure 6.2, the simulator vendor will supply you with database access routines to extract relevant data directly from the simulator's internal data structures. We still must write the translation program, and errors in the application of the ATE will likely be found through tedious debugging on the ATE system.

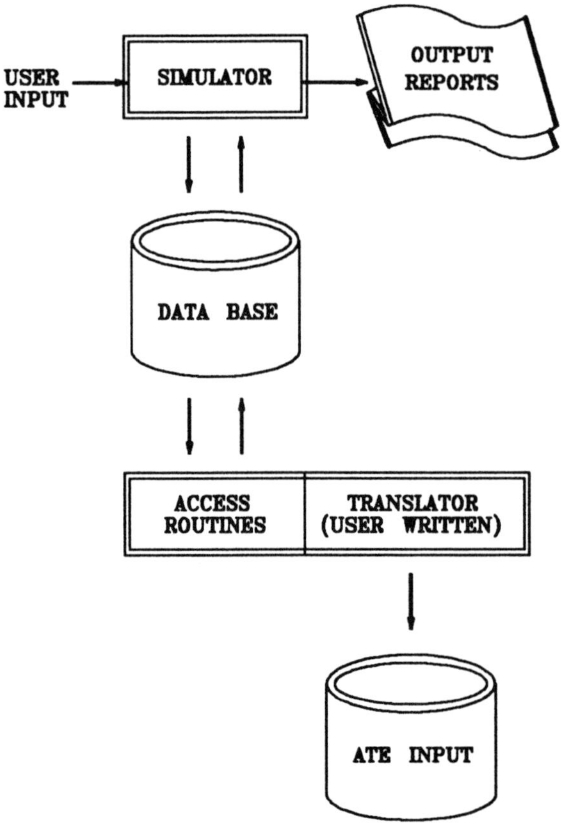

Figure 6.2

In a more advanced approach, here termed an 'expert' translator, the translator is a major system itself. It accepts a description of the ATE system's capabilities and then examines the simulation output to 'figure out' how to use these resources (Figure 6.3). This requires detailed understanding of the simulator output, the input syntax of the ATE system, and how the output waveforms can be manipulated into ATE input within the rules of its operation. This is a good deal more complicated than just reformatting a sea of 1s and 0s. Therefore, such an approach must be specifically tailored. Some third party software suppliers have taken this approach. An advantage of this method is that the software can check for violations of the ATE system capabilities and flag problems before the user gets on the ATE system.

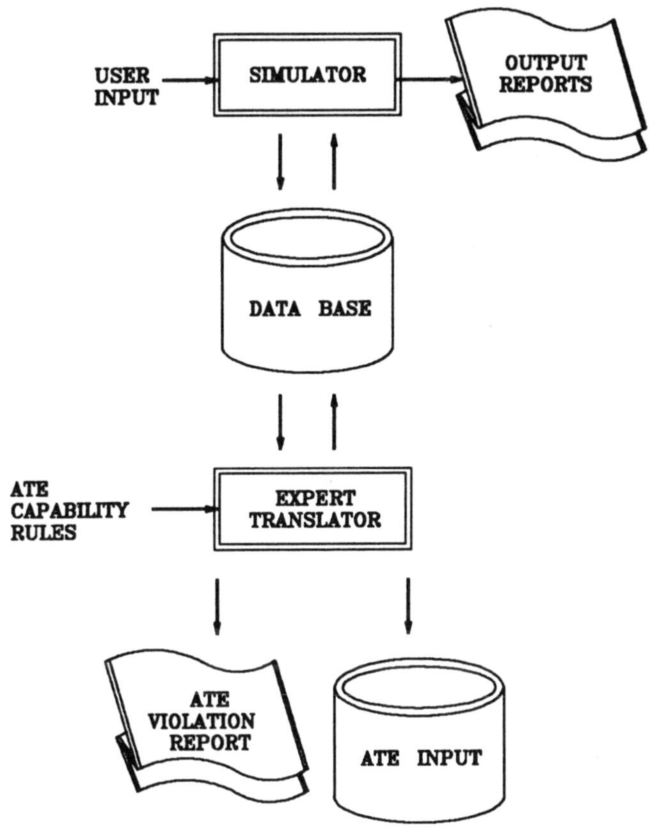

Figure 6.3

Perhaps an ultimate solution, here termed an 'expert test development' system, is a tightly integrated system that is 'aware' of the needs of the test development process (as stated in earlier chapters of this book), is 'aware' of the capabilities of the ATE system, and guides the user through the test development process. (See Figure 6.4.) It prevents wasted simulation by watching for mismatches between the simulator capabilities and the ATE system. Input to the simulation is checked for compatibility with the ATE system's driver capabilities. When simulation is complete, the simulated circuit behavior is checked to see if the ATE system will have any difficulty monitoring the simulated response. The output files produced are already in the form suited to the ATE system.

Note that several systems may still be used to create a test. For example, the design of a board may be done on this ideal system, but some IC designs may be done on another system because of, for example, restrictions imposed by the IC foundry. The overall test development process needs to be flexible.

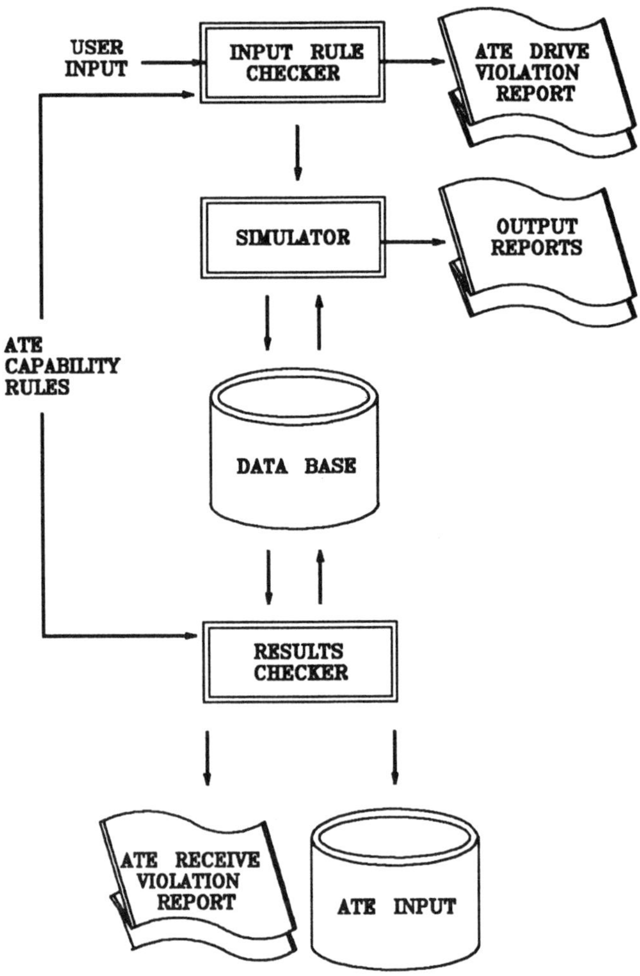

Figure 6.4

6.2. Qualification of Data

A simulation system has all (or nearly all) the data needed to create a working digital test on an ATE system. However, the data required may be much more than just vectors. They may include internal node states and circuit topology. Further, they are often 'logical' data that are not precisely matched to the physical implementation of the circuit. Appropriate fixes will be needed. Finally, incompatibilities in character sets or symbology may occur. For example, the simulator may name active–low signals with trailing dashes (like CLEAR–) but the dash is not legal on the ATE system. The ATE system may use ASCII character sets, but the simulator uses something else. Signal names may be up to 16 characters on the simulation system but limited to 12 on the ATE. Many of these problems can be intercepted and fixed during translation.

Another extremely useful item to include in *all* translated results is a heading that appears to the ATE system as simple comments but contains date, time, and data tracking information. This will allow the ATE user to identify the source of the data and to differentiate among revisions and multiple translations. Without such information, confusion may result.

6.3. Bidirectional Interfaces

When simulating the behavior of a circuit that has bidirectional I/O pins, the creator of vectors has to be aware of the direction of data flow. It may be the case that the output of the simulation is a record of the state of these I/O pins, but without the direction information. How, then, do we translate this information into ATE tests where bidirectional pins must be properly driven?

We might attempt to correlate the simulation output with the original input to determine when the pin states are a result of circuit outputs. This may be the only choice if the simulation cannot be redone. However, if one is still in the planning stage, there is a simple technique to make the translation problem for bidirectional pins much easier. This technique relies on the creator of the simulation model to identify each bidirectional pin or group of related pins (like buses) and assign a companion 'dummy' pin, which is unconnected to anything in the circuit. This pin is the 'direction' signal used to indicate the direction that data are flowing on the pin or pin group. Define '1' to mean that the circuit is being driven and '0' to mean that data are being received from the circuit. Then, whenever vectors are fed to the simulator, set the appropriate states on the dummy pins. This can be taken care of automatically if the simulator input language has a 'macro' or cycle definition capability. The details of managing dummy signals can be specified once in the macro definition. When it comes time to translate the data for the bidirectional pins, the associated dummy signals contain a record of direction information. This information is stripped out during translation, removing the data for nonexistent dummy pins and also producing the correct drive/receive states on the tester channels.

6.4. Manual Intervention

If the capabilities of a test system are properly addressed during test preparation, then ideally the data from the simulation can be translated directly into an ATE test with no need for changes. In reality, it is possible that the resulting tests will need modification before they will run on the test system. This has two primary causes: discrepancies between the logical and physical models of the circuit and initialization.

If there are discrepancies (usually at board–level test), they probably were introduced deliberately to overcome or ease some problem with the simulation. If this was not thought out in advance, then an ATE debugging session of legendary duration may result.

On the other hand, initialization is a fairly common problem. Again, owing to difficulties in simulation, an initial state may be assumed to exist when the first vectors are applied. Since that initial state will not exist in reality, a manually generated homing sequence will have to be added to the test as a preamble that produces the proper starting state. This is not necessarily difficult, unless the initial state chosen as the start state for simulation is difficult to target with a homing sequence. A homing sequence will require access by a number of receiver channels to key observation points within the circuit. If these are not available, or if for any reason the homing sequence cannot be written, we may have to discard the entire simulation effort. Unfortunately, creating solid initializations is more of an art than a science.

6.5. Debug

When it comes time to run the test on the ATE system, we hope to enjoy the fruits of past labors. However, there may still be difficulties. The ATE system will certainly have a set of tools to help us figure out what the problem is, and maybe it can be patched up on the system (and carefully documented, of course). However, we may be forced to return to the simulation system to fix the problem, requiring new simulation effort and a new translation. It is therefore highly recommended that we try out a very small set of vectors on the ATE system to see if they, the circuit model, and required initialization are going to work. We should do this *before* investing hours of effort in producing a complete test. If the circuit is being developed in parallel with the test, then such trials will not be possible since hardware does not exist yet. However, the strategy for test should be developed as early as possible so that testability enhancements can be made in the design, particularly those that aid the initialization effort. The value of such enhancements quickly becomes evident when the first simulations are attempted.

6.6. The Physical Link

After data have been suitably prepared on our CAE system, it must be moved to the ATE system. In a tightly integrated CAE/ATE solution, this is a simple matter. However, if this is not the case, or if more than one simulator system has been employed in test development, then part of our planning must include the physical link issue.

The physical link may be active or passive. An active link is a direct connection between the two machines: Simple examples are RS–232 or an 802.3 LAN. A passive link is accomplished with a magnetic storage media such as magnetic tape or floppy disc.

The physical location of the two systems may dictate the approach taken. If they are relatively close (less than 100 meters) then direct RS–232, 802.3

LAN, or other high–bandwidth electronic links are candidates. If the systems are many miles apart, then RS–232 transmission over telephone lines may be the answer; being cognizant of data volume versus transmission speed and reliability. It may turn out that magnetic tape is a better match. For instance, if a large simulation creates a 1 Mbyte result, then nearly 2 hours of 1200 baud transmit time will be required.

The type of data involved may also influence the choice of link. For example, if the ATE system accepts data in ASCII (or some other character set) form, then RS–232 may be quite appropriate. However, if some of the data are in a binary form, then RS–232 may not work since the binary data may contain codes or parity bits incompatible with the RS–232 protocol. Magnetic tape or disc offers the greatest flexibility here, primarily owing to its maturity as a data porting technology. Data densities of magnetic media are also high, although the transport mechanisms (tape drives, etc.) can be quite expensive and must exist on both systems as well as compatible read/write utilities.

Chapter 7
Conclusion

Chapter 7

Conclusion

Throughout this book, we have been examining ways that we can use our CAE and ATE investments together more effectively. The combination of these resources can offer more than just the sum of their components.

Such synergism occurs in several ways. If CAE and ATE tools are effectively combined, then the personnel that use them will receive common training, speak with a common vocabulary, and share in common goals. Their managers will notice that it is easier to make tradeoffs, particularly those crossing departmental boundaries. It will be easier to gain cooperation between design and test.

Because design and test people are more closely allied and better understand each other, it will be easier to quantize the value of the many design for testability techniques that may be applied. In any given design situation there are many different choices to be made on how to 'add testability.' When the design effort is integrated with the testing effort, it is much more obvious which choices are best. Similarly, rather than having a test department issue testability guidelines by fiat, they can now choose guidelines in a fashion more sensitive to the constraints imposed on the designers.

The future holds great promise for improving both the design and test efforts. As they become more closely integrated, the two experience bases they have will also become one. The past has shown that perhaps the greatest synergisms occur when separate experiences can be combined.

INTEGRATING DESIGN AND TEST: USING CAE TOOLS FOR ATE PROGRAMMING

Appendix A
Waveform Quality and Tester Specifications

Appendix A

Waveform Quality and Tester Specifications

In the foregoing chapters of this book, especially Chapter 2, it has been noted that waveform quality can be an important issue when designing a test. This quality is subject to influences determined in part by the ATE system, the fixturing, the device under test, the laws of physics, the approach to testing, and the user's understanding of the problems. In this appendix, examples will be shown of how these various influences must be addressed, particularly since the successful use of CAE tools may otherwise be compromised.

There are three areas of concern.

1. Waveform quality: Is the ATE system producing valid logic levels?
2. Degradation of tester specifications: Are the quoted ATE specifications realizable in a given application?
3. Safety issues: Is there a danger that signal conditions may exist that could cause damage to the circuit under test?

A.1. Waveform Quality

Consider Figure A.1, which shows an ATE driver connected to the input of the device under test through fixturing.

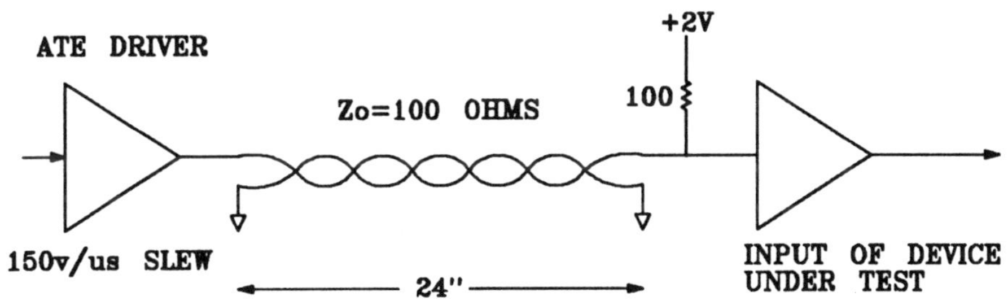

Figure A.1

This case is quite representative of a functional board test setup. If it were an IC tester, the fixturing might be done with printed circuit striplines of only 8 to 10 inches and the driver slew rate might be 1000 v/usec. Figure A.2 shows what the signal actually looks like as seen by the device under test.

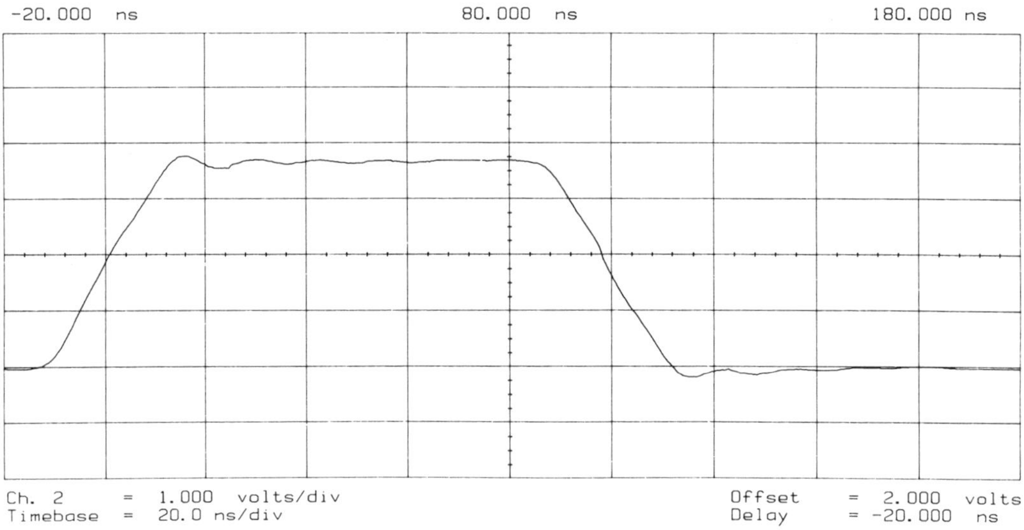

Figure A.2*

The waveform shown is a 100 nsec TTL compatible pulse between 0 and 3.5 volts. Because the driver rise/fall time is determined by the 3.5 volt swing and the slew rate of 150 v/usec, it takes over 20 nsec for the transition to occur. There is almost no over shoot/under shoot. This is an example of a high quality waveform. Since the device under test is a very low load receiver, it is easy to attain this level of quality consistently. Causes of quality degradation would likely be caused by improper matching of line impedance Zo with the termination resistor, lack of termination, or not providing adequate signal ground return paths.

When using the same setup to do in–circuit testing, significant loading is now imposed at the destination of the signal, and this loading is often both active and nonlinear in its response to an overdriving pulse. Consider the case where we are overdriving a simple LSTTL driver that is in the zero state. By using the same waveform and levels as before, the output of the LSTTL driver is forced to logical 1. Here, the circuit of Figure A.1 has been changed, removing the 100 ohm termination resistor and adding in its place an LSTTL driver which is driving a 0 onto the node, as in Figure A.3a. The resulting waveform is shown in Figure A.3b.

* The oscillographs used in this appendix were provided by Phil King of Hewlett–Packard Company.

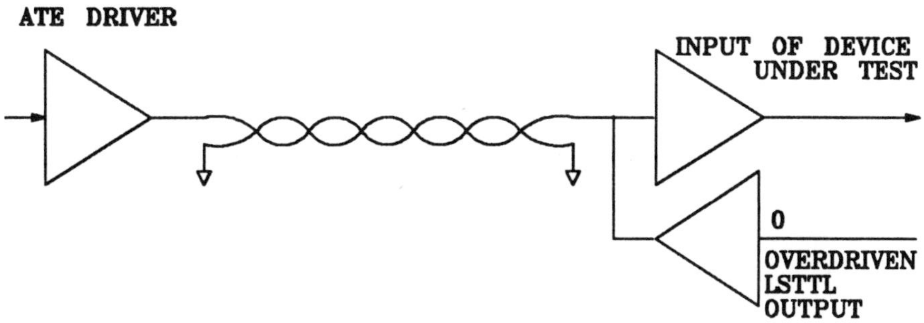

Figure A.3a

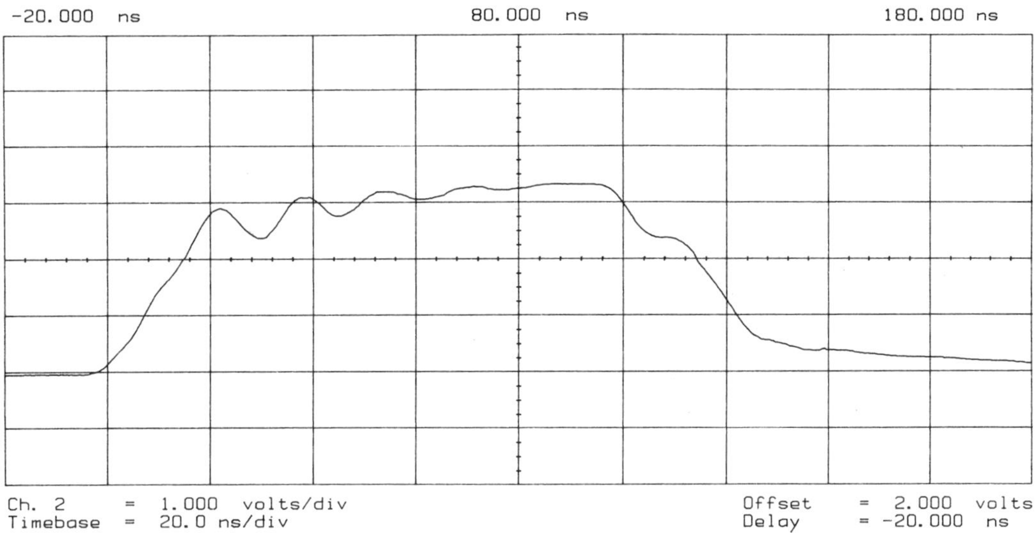

Figure A.3b

Notice first that there is some ringing caused by transmission line effects. This amount of ringing is not serious. Notice that the signal finally stabilizes at a level about 200 mv below the unloaded waveform, owing to the current flow through the source impedance. During the falling edge, there is some sluggishness. This is because of the time it takes to discharge the transmission line. The waveform overall has not been very seriously degraded. This is because the LSTTL driver is not presenting a heavy load. In other situations, the degree of degradation can be completely unacceptable and, as we shall see, nearly unbelievable phenomena may occur.

Consider the situation where the LSTTL driver is replaced by a high speed TTL driver. The IC in question is actually implemented internally in ECL. The output stage must perform a level translation and amplification back to TTL levels. To enhance speed, this output stage uses positive feedback. This causes problems when we try to pull the driver high. The information shown in Figure A.4 is the result of a 200 nsec pulse applied by our setup of

Figure A.1, superimposed on the reaction of the IC when being overdriven to 1. (Note the scale changes of 50 nsec/div and 2 v/div.) The overdriven point oscillates at about 18–MHz, with a voltage swing greater than 12 volts!

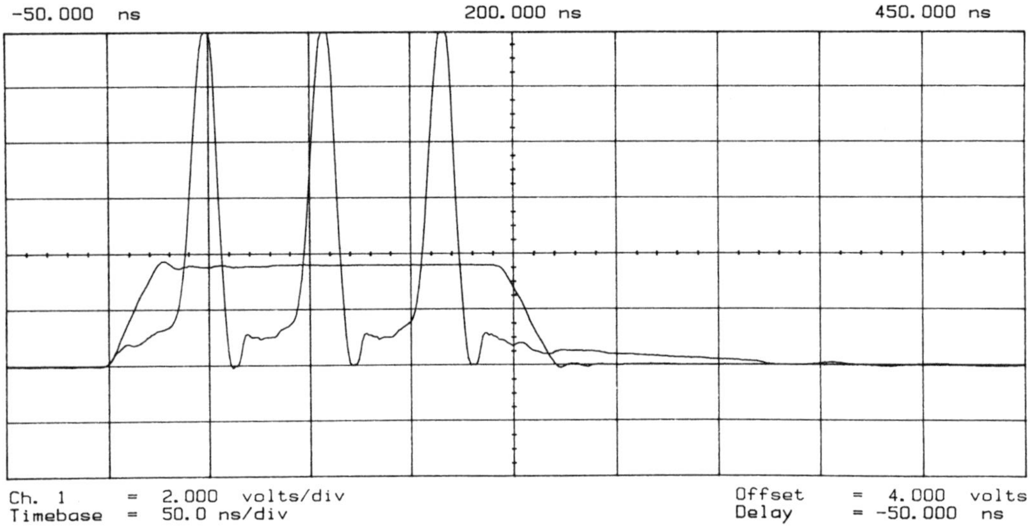

Figure A.4

Lest the accusation be made that faulty fixturing or ATE driver design is to blame, we completely replace the driving circuitry with the ultimate DC driver shown in Figure A.5.

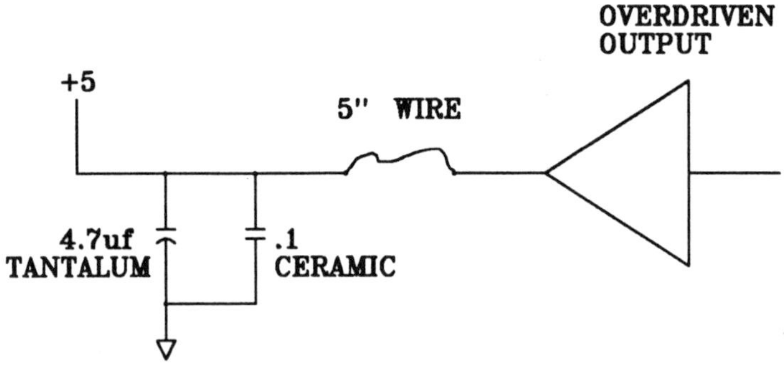

Figure A.5

This DC driver is only 5 inches away from the overdriven output and does not oscillate itself. However, Figure A.6 shows the waveform that exists at the point of overdrive. (Again note scale changes.)

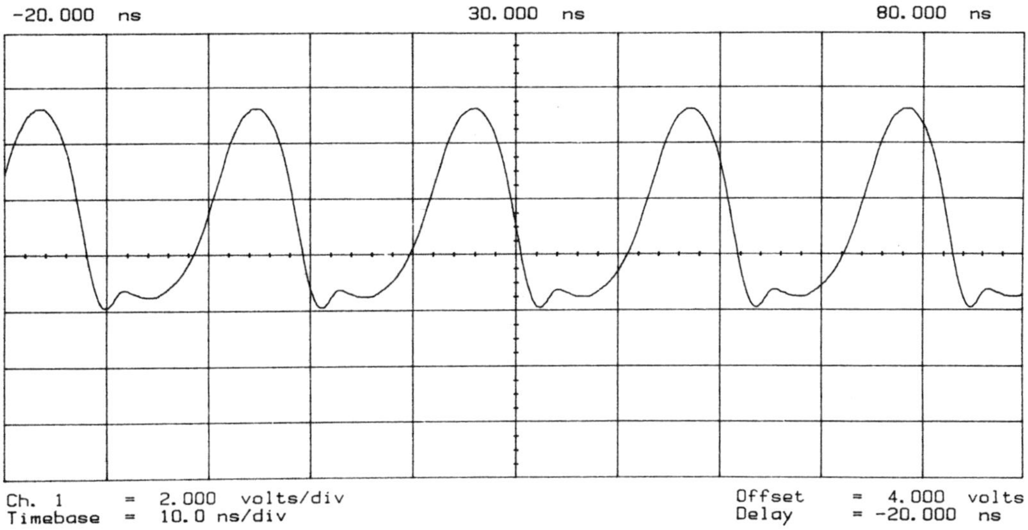

Figure A.6

Again oscillation occurs, this time between 2 and 9 volts and about 45–MHz. With 3 inches of wire, oscillation still occurs, albeit at higher frequency and lower amplitude.

Now the user of a CAE system preparing a digital in–circuit test may never anticipate such a problem. However, if it occurs, there is no physical cure such as getting a better fixture or ATE system. The simplest solution is not to overdrive this device when it is in the 0 state. This means we need vectors to put the device into a 1 state where no problems occur. The user of the CAE system can easily provide these. This is an illustration of how CAE systems may be used after the original design to serve the needs of the test department.

A.2. Degradation of Tester Specifications

When preparing an ATE test, one should examine the timing specifications of the ATE system to see what limits to observe when doing a simulation. However, sometimes the loading presented by the fixturing and/or the circuit under test must also be taken into account because they have the effect of loosening the original state specifications to a degree that can be surprising. If we intend to press hard on the limits of our ATE system, we need to first consider how its specifications may degrade before investing time in simulation.

We have already seen with our study of the LSTTL overdrive situation that the loading presented by our circuit can begin to change the wave shape of our overdriving signal. As might be guessed, as the load increases, additional changes occur. Consider the unloaded 200 nsec TTL driver pulse in Figure A.7 as a reference in the same overdrive setup used before. Note the transition time from 0 to 2 volts (logical 0 to 1) is about 10 nsec. The transition back to 0 begins at time 190. Once below the 0.6 volt level, the 1 to 0 transition is complete. This takes about 25 nsec.

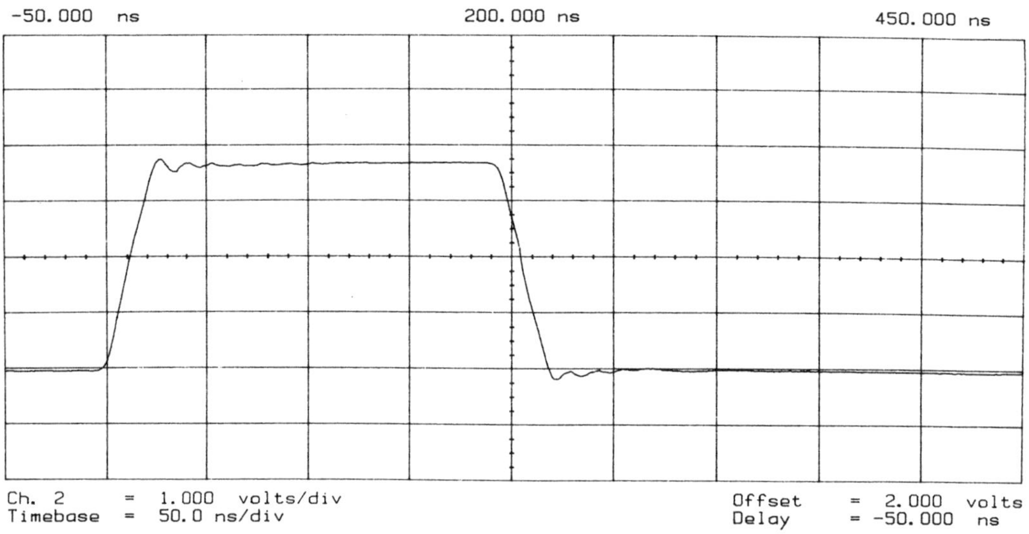

Figure A.7

Now we use this driver to overdrive (from 0 to a 1) the output of a FAST™ TTL gate. The test setup is identical to Figure A.3a. The resulting waveform is shown in Figure A.8. While this is a more powerful device than an LSTTL device, it is representative of FAST™ devices in general. In other words, it is not a special case such as a high powered line driver.

™ FAST is a trademark of Fairchild Company.

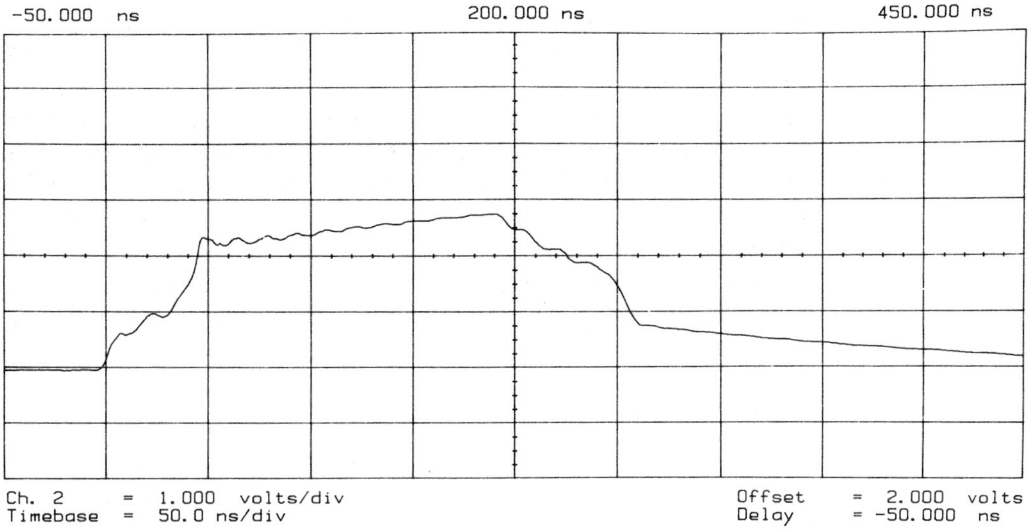

Figure A.8

The transition time from 0 to 2 volts is not 10 nsec any more, but 45 nsec. The highest voltage attained is 2.75 volts. The downward transition starts at time 190 but does not cross the 0.6 volt level until about 110 nsec later. The net effect is that the waveform can be interpreted logically as having been delayed 35 nsec and made wider by about 60 nsec. This certainly represents a significant change.

Now consider the waveform in Figure A.9, which is the result when overdriving a FACT™ TTL compatible CMOS gate. This is a recently introduced high performance CMOS device. Again, compare the resulting waveform to the ideal waveform of Figure A.7.

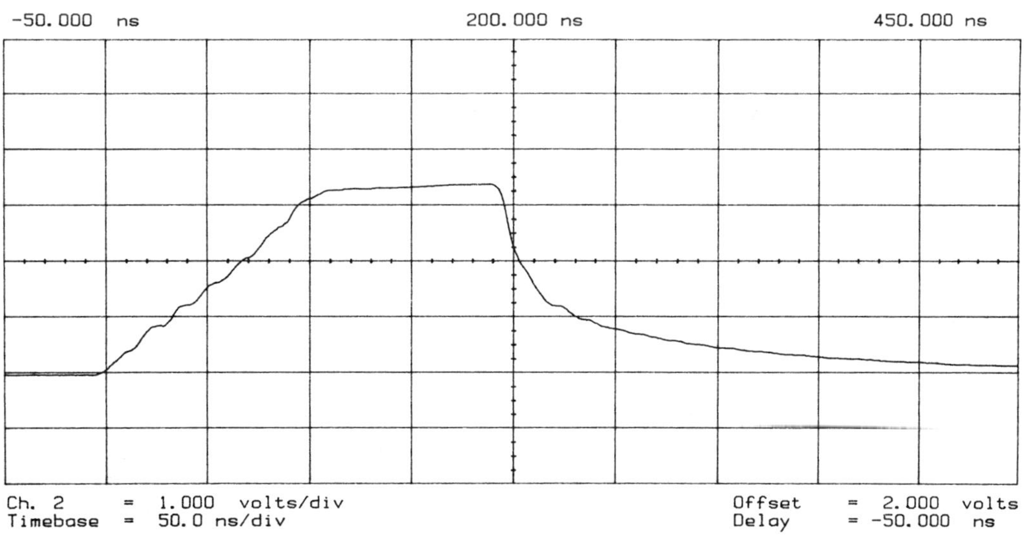

Figure A.9

™ **FACT is a trademark of Fairchild Company.**

The rising transition shows a 'staircase' pattern caused by transmission line effects. The transition time from 0 to 2 volts is not 10 nsec but now is 67 nsec. The highest voltage attained is only about 200 mv lower than the unloaded case. The downward transition starts at time 190 but does not cross 0.6 volts until about 90 nsec later. The net effect is that the waveform can be interpreted logically as having been delayed 57 nsec and made wider by about 20 nsec. This is a significant change and quite different from the preceding case.

The net result of this is that the higher powered, faster devices may have to be tested more slowly because of signal degradation, an interesting paradox.

A.3. Safety Issues

In this section, we consider the case where the in–circuit technique is used to overdrive a low–power (or higher impedance) load such as a low–power CMOS device. Again, our experimental setup is the same as before, using 100 nsec pulses. In Figure A.10, we see the resulting waveform shown with a 20 nsec/div timebase.

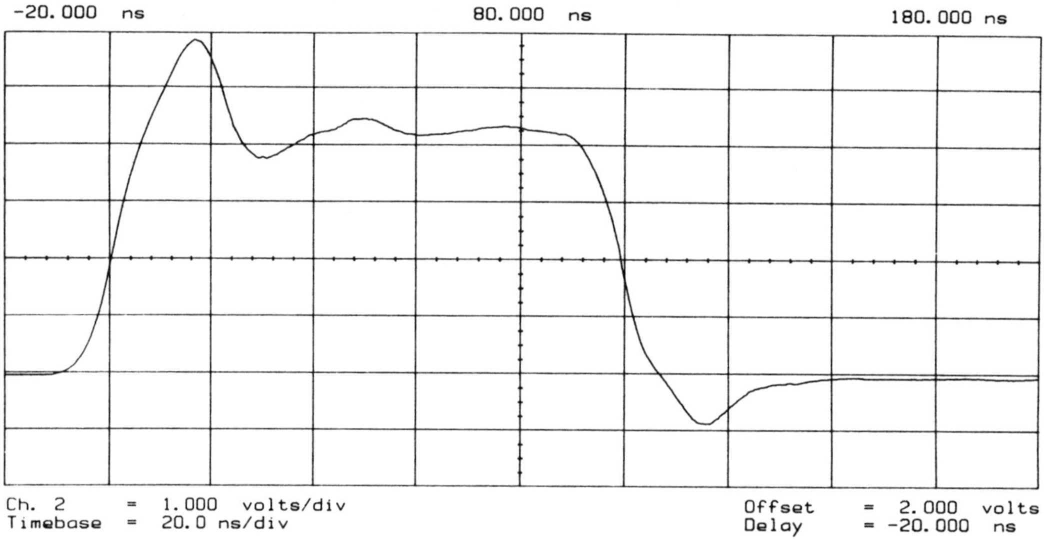

Figure A.10

Note that there is significant overshoot and undershoot now occurring, owing to the mismatch of the load to the transmission line. Indeed the peak–to–peak values are about −1 to +6 volts. If the CMOS device is operating at normal +5 volt power levels, the fact that the these over shoots/under shoots occur could excite latchup failures in the inputs of other devices connected to the node. Imagine what might happen to fragile devices on a node connected to the oscillating driver seen earlier in Figure A.4!

Now, to solve the problem of over shoots/under shoots during backdriving in this situation, we must control the slew rate of the driver. In other words, we deliberately slow down the driver (degrade its specifications) to allow the transmission line to track the transition of the driver. The example of Figure A.10 used a slew rate of 150 v/usec, but the example shown in Figure A.11 uses a slew of only 40 v/usec. Note the scale change in the time base.

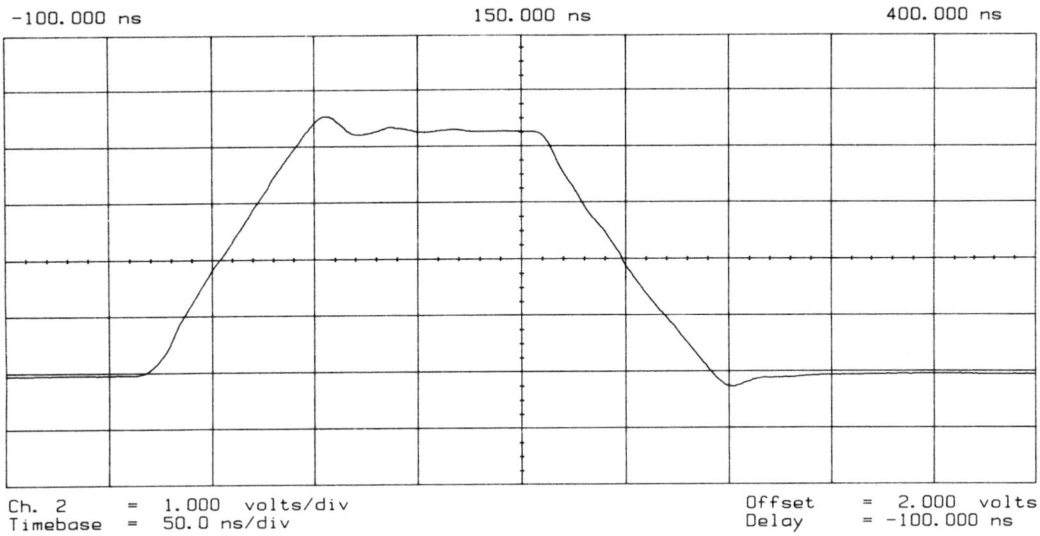

Figure A.11

Now see that the amount of over shoot/under shoot is quite small. However, if the need for slower slew rate had not been anticipated during the preparation of the test on your CAE system, we may have to discard some of our work and downgrade the speed of the simulated test.

A.4. Summary

This appendix has shown some common problems that may occur when testing a circuit. In general, the devices to be tested are orders of magnitude smaller than the testers themselves. This size imbalance drives the requirement for fixturing that introduces extra path length and loading into the stimulus/response paths. While ATE systems are designed to minimize the introduction of these spurious factors, they still are present. Especially as test speeds go higher, test development must account for these factors.

INTEGRATING DESIGN AND TEST: USING CAE TOOLS FOR ATE PROGRAMMING

Appendix B
An ATE Checklist

Appendix B

An ATE Checklist

This appendix contains a checklist of questions that CAE users' should ask of the ATE department (and themselves) to aid in the development of ATE programs. Since the elements of the checklist are already defined mainly in Chapter 2, the questions will be brief. The primary contribution of this chapter is to enumerate the common questions and point out how some questions are interrelated.

B.1. The Type of Test

The first major area of definition concerns the type of testing approach to be used. This will influence how we use our CAE tools. Note that a combination of techniques may be used as we partition a large testing problem into manageable pieces.

The definitions given next will contain overlap. They are meant to be self–contained.

B.1.1. In–Circuit

Is the circuit to be tested going to be approached with the in–circuit test technique? If so, ask the following questions about the circuit to be tested followed by another set of questions about the ATE system:

About the circuit to be tested:

- What logic family is it?
- How many I/O pins does the device have?
- How many additional channels are needed for disabling?
- What are its overdrive requirements?
- Is it stable during overdrive?
- What are its disable vectors?
- What load will it present to the tester?
- What load can it drive toward the tester?
- Is fixturing going to be straightforward?
- Are any of its pins constrained in its application?
- Is it synchronizable *as constrained*?
- Is a homing sequence going to be needed *as constrained*?
- What faults are the tests supposed to detect?
- What is the estimated number of vectors required?
- Could formatting reduce the effective vector count?
- Could formatting reduce the effective vector rate?
- Will algorithmic vectors (e.g., RAM tests) be required?

About the ATE system:

- How many parallel channels can run simultaneously?
- What overdrive capability exists?
- What load does a channel present to the circuit?
- What loading effects will the fixturing present?
- How will driver transition times degrade?
- What will be the resulting pin–to–pin skews?
- What restrictions are there on comparator windows?
- What is the effective bandwidth of a channel?
- What is the maximum sequencing rate of the tester?
- What is the sequencing jitter?
- What is the depth of vectors supported?
- What algorithmic pattern generation is supported?
- How does the tester handle initialization?
- What (if any) formatting capability is supported?
- How many channels can be formatted?

B.1.2. Functional Cluster

Is the circuit to be tested a cluster of ICs for which the in–circuit technique will be used to test them as a group? If so, you should ask the following questions, about the cluster, and about the ATE system.

About the cluster:

- What logic families are used in the cluster or in upstream devices?
- How many I/O pins does the cluster have?
- How many additional channels are needed for disabling?
- What are the overdrive requirements?
- Will overdriving experience stability problems?
- What are the required disabling vectors?
- What load will be presented to the tester?
- What load will the tester represent to the circuit?
- Is fixturing going to be straightforward?
- Is the circuit synchronizable?
- Is a homing sequence going to be required?
- What faults are the tests supposed to detect?
- What is the estimated number of vectors required?
- Could formatting reduce the effective vector count?
- Could formatting reduce the effective vector rate?
- Will algorithmic vectors (e.g., RAM tests) be required?
- Will fault isolation data be required of internal nodes?

About the ATE system:

- How many parallel channels can run simultaneously?
- What overdrive capability exists?
- Can several different logic families be tested together?
- What load does a channel present to the circuit?
- What loading effects will fixturing present?
- How will driver transition times degrade?
- What will be the resulting pin–to–pin skew?
- What restrictions are there on comparator windows?
- What is the effective bandwidth of a channel?
- What is the maximum sequencing rate of the tester?
- What is the sequencing jitter?
- What is the depth of vectors supported?
- What algorithmic pattern generation is supported?
- How does the tester handle initialization?
- What (if any) formatting capability is supported?
- How many channels can be formatted?
- Is guided probe fault isolation supported?
- What data is required for the guided probe?

B.1.3. Functional Edge Connector

Is the circuit under test going to be approached with an edge connector functional test technique? (This includes board level testing, IC wafer testing, and IC package testing, although only board level testing will be concerned with fault isolation data.) If so, ask the following questions about the circuit to be tested and the ATE system.

About the circuit to be tested:

- How many I/O pins does the circuit have?
- What load will it present to the tester?
- What load will the tester present to the circuit?
- Is fixturing going to be straightforward?
- Does the circuit contain internal oscillators?
- At what frequency do these oscillators run?
- Does the circuit contain asynchronous subsystems?
- Is the circuit synchronizable?
- Is a homing sequence going to be required?
- What faults are the tests supposed to detect?
- What is the estimated number of vectors required?
- Could formatting reduce the effective vector count?
- Could formatting reduce the effective vector rate?
- Will algorithmic vectors (e.g., RAM tests) be required?
- Will fault isolation data be required?

About the ATE system:

- How many parallel channels can run simultaneously?
- What load does a channel present to the circuit?
- What loading effects will the fixturing present?
- How will driver transistion times degrade?
- What will be the resulting pin–to–pin skews?
- What restrictions are there on comparator windows?
- What is the effective bandwidth of a channel?
- What is the maximum sequencing rate of the tester?
- What is the sequencing jitter?
- What is the depth of vectors supported?
- What algorithmic pattern generation is supported?
- How does the tester handle initialization?
- What (if any) formatting capability is supported?
- How many channels can be formatted?
- Can the tester be phase locked to an on–board oscillator?
- Can vector application be interrupted by handshake signals?
- What data is required for guided probing?

B.1.4. Special Considerations

When developing a digital test at any level, there may be need for some special considerations. Here are some examples.

The circuit under test will be supplied with power. How much? Does the tester have enough capacity? Will cooling be necessary? Are several different levels required? What amount of regulation is required? Would a power supply short be dangerous?

Is analog circuitry present within the circuit under test? What effects will it have on the digital test? Will it cause our logical model to differ from the actual circuit? Will tester timing specifications be altered?

Do on–board oscillators exist? Can they be disabled? Can tester controlled clocks be substituted? Must we synchronize and phase lock the tester to an oscillator? What would be the effects on timing specifications of the tester? How would we diagnose the cause of failure when phase lock is not achieved?

Is there an asynchronous subsystem? Can we use tester vector synchronization to track asynchronous handshakes? How will we 'fool' the simulator into creating data for this circuit? Could we avoid all these problems by treating the subsystem separately as a cluster?

Are there features of the design deliberately added to improve testability? Can we use them at this level of test? Were they added to improve testability at a different level of test? Were they added to address the same fault spectrum we are currently interested in? Will they improve testability but cause unusual demands to be made of the tester architecture? As an example, a scan test design discipline (such as LSSD) may cause excessive serialization of a test. This can result in excessive run times or exhaustion of the tester vector storage capacity on a tester not built for LSSD.

B.2. Hardware Checklist

The following section gives a list of parameters relevant to ATE hardware. Some may be irrelevant in individual situations; for example, backdrive current capacity with respect to functional IC testing. Again, the descriptions of the resources, their capabilities, and their limits were itemized in Chapter 2.

B.2.1. Resource Count

Take an inventory of the resources available from the ATE system as configured and the maximum possible. This will tell us if our testing problem will fit on the system and, if not, whether more resources can be added to cover our situation. Obviously, if enough cannot be added, we must change our approach to testing our circuit.

- How many drive/compare channels (before/after multiplexing)?
- How many channels are formatted?
- Is there external clock and vector synchronization?
- How many device power supplies are supported?
- Are there algorithmic and/or random vector sequencers?

B.2.2. Channel Technology

This section lists some specifications made of the replicated drive/compare channels. Note that some will also contain programmable loads as well. Be especially aware that specifications are subject to degradation due to loading and fixturing as covered in Appendix A. Implicit in a number of these questions are concerns about signal quality.

- For overdriving, what it the driver current capacity?
- What is the driver rise/fall time (loaded and unloaded)?
- What is the minimum pulse width of drivers?
- What is the driver pin–to–pin skew?
- Is there a channel impedance specification?
- What is the channel bandwidth?
- Is there a minimum drive to receive time?
- Is the comparator a sampling or window comparator?
- What is the minimum comparator window?
- Are there single or dual comparator thresholds?
- What is the comparator pin–to–pin skew?
- If there are programmable loads, what are the source/sink currents and voltage threshold?

B.2.3. Fixturing

All testers must make physical contact with the device under test, so all have some sort of custom interface or fixture. A prime concern with fixturing is how it affects the quality of the signals sent to and received from the device under test. The following questions are related to this.

- What is the path impedance (if controlled)?
- What is the typical signal path length?
- What is the typical capacitive load?
- Are signal return paths dedicated (e.g., twisted pair) or common?
- What are the termination options?

B.2.4. Formatting Capabilities

Within a vector cycle, drivers may be capable of producing multiple edges and receivers capable of multiple windows if formatting capability is provided. Assume that the formatting capability modulates the data being applied to the device under test. That is, one bit per vector may be applied to the device with formatting producing several edges. See the discussion in Chapter 2.

- How many channels are formattable?
- How many control edges (or events) per cycle?
- How close together can these edges be?
- Are edges freely assignable to driver and comparator?
- Are there restrictions on activities near period boundaries?
- Are there restrictions on activities across period boundaries?
- How many different cycles can be defined?

B.2.5. Vector Limits

Vector (pattern) data can take many forms and may refer to simple 1s and 0s or to the highly formatted waveforms of the previous section. In any event, these must be stored within the tester, usually in RAM within each channel.

- How many vectors can be stored?
- Is any vector compression scheme used to enhance this?
- What is the effective vector depth?
- Is any unusual storage technique used for special applications such as scan testing?

B.2.6. Sequencing Limits

Along with the storage of vectors, their proper sequencing is important. There are two issues here: support for linguistic structures such as loops and subroutines, and support for vector compression schemes to enhance the apparent depth of the vector storage.

Advanced linguistic structures are often included in ATE vector languages for the convenience of a human author of vectors. When translating from a CAE system, the existence of loop or subroutine structures may not be obvious in the 'sea of bits,' and it will consume processing time to search for them. Most translation efforts will end in a linear vector–for–vector result.

Vector compression schemes are often hidden from the user of the ATE system. In essence, the compiler of the ATE vector language looks for compressible items while doing the compilation. The resulting level of compression may be difficult to predict beforehand.

- Does the vector sequencer support loops?
- Are there limits on the size of the loop or repeat count?
- May loops be nested?
- Does the sequencer support subroutines?
- May subroutines be nested?
- Do subroutines support parameter passing?
- Will the sequencer inject jitter into pattern application?

B.2.7. Sequencing Speed

This is a simple question, with some interesting nuances.

- How quickly can the sequencer produce vectors?
- How slowly can the sequencer produce vectors?
- What is the resolution on sequencer speed?
- What is the accuracy of sequencer speed?
- Can vector periods be changed on the fly?
- What is the dynamic range on period changes?
- Is the sequencer delayed during certain operations, like exiting loops, subroutines, or monitoring homing sequences?
- Can the sequencer be phase locked to an external period source?

B.2.8. Initialization Capabilities

The initialization problem may not affect the ATE system or the sequencer at all if a simple synchronizing sequence exists for the circuit under test. However, if the circuit requires a homing sequence, then support must come from the sequencer.

This support comes in the form of the ability to reinterpret the meaning of the comparator 'fail' lines. Instead of fail, they are interpreted as 'absence of a match.' Here, the comparators are being used to watch for a particular state to occur. During this time, conditional branching is necessary to alter the input sequence based on whether or not the comparators observe a match. Until the proper initial state is achieved, the test cannot proceed. There is a danger that a failure will prevent this state from ever occurring. Hence, the possibility of an infinite loop must be addressed.

Here are some questions:

- Does the sequencer support homing sequences?
- Are there restrictions on which (or how many) comparators may be used?
- During the homing process, are there speed restrictions on the sequencer?

- Are there any dead cycles during the homing process? (For example, are there dead cycles on conditional branches or pipeline flushes that occur when homing completes?)
- How does one avoid the infinite loop problem?
- Does the sequencer hold off the vector counter during homing?
- Do the comparators switch from match mode to fail mode automatically when homing completes?

When a homing sequence is necessary before testing proceeds, it will be necessary to insert the vectors and control sequencing for the sequence before the test vectors. This brings up additional questions.

- Is the input to the ATE system able to be edited, so that the homing sequence can be added?
- If not, how does one specify the homing sequence to the ATE system?

B.2.9. Synchronization Capabilities

When addressing circuits with asynchronous subsystems, find out if it is possible to remove the effects of asynchronous interactions by having the tester monitor key handshake signals and interrupt the sequencing flow until key events occur.

- Does the ATE system possess synchronization features?
- Can a vector be interrupted between stimulus and response phases?
- How many channels have this hold off capability?
- What if the awaited event never occurs?

B.2.10. Safety

How safe is the device under test during testing? It is possible to do damage to the circuit while attempting to test it. In recognition of this, some testers have hardware and/or software features that serve to protect the device during testing. The damage issue is of obvious concern when using in–circuit overdrive as a test methodology. However, it can also be an issue during functional testing, (1) because signal quality problems may cause improper levels to be applied to a device (see Appendix A) and (2) one may lose synchronization with a circuit's bidirectional signals and begin to 'fight' with them. This loss of synchronization may be the result of incomplete or inaccurate initialization/homing.

Here are some questions:

During in–circuit testing:

- Does the ATE system limit the time that overdrive may be done? (This may cause you to shorten or segment your tests.)
- Does the ATE system allow the selection of driver edge rates? (See the example in Appendix A.)
- If edge rates are selectable, which rate and attendant set of specifications must you use?
- Does the tester have an unpowered shorts testing capability to eliminate threatening shorts before digital testing?

During functional testing:

- Does the ATE system sense the fact that bus contention is occurring and stop the test?

B.3. Software Checklist

This section gives a list of parameters relevant to ATE software.

B.3.1. Character Set Limitations

The programming on an ATE system is done within the character set convention of its controller. Since the CAE system may have a different controller, some conflicts may exist between representation of data on the two systems.

- What character set convention (e.g., ASCII, EBCDIC) is used on the ATE system?
- What special characters (e.g., ESCAPE) are allowed or disallowed on the ATE system?
- If an electronic link (such as RS–232) is used to transfer data to the ATE system, will the transmitted characters be legal within its protocol?

B.3.2. Pin/Node Identifiers

When transfering a digital test to an ATE system, 1s and 0s are only part of the data. Also included are the names of signals, nodes, and pins.

- What are the length restrictions on names?
- What characters are legal in names?

B.3.3. Topology

For functional testing using guided probing, it will be necessary to provide the ATE system with a topological description of the circuit, and this should be readily available within the CAE system database. These data are made up of signal, node, and pin names as well as the parts list. The same questions directly above are relevant and the answers may vary depending on what is being described. For example, signal names may not be allowed to start with a numeric digit, but part designator names may have less restrictive rules.

B.3.4. Vectors, Formats, and Cycles

How are vectors, waveform formats, and tester cycles expressed in the ATE system language? Formats and cycles may have complicated syntaxes. Vectors may be quite lengthy, at least in that they may well exceed 80 columns of data and become unwieldy on CRT screens.

- How are vectors expressed, in particular, when their descriptions may require more than one line of text?
- How are formats and cycles expressed?

B.3.5. Internal States

To support functional test fault isolation by using a guided probe, one may need to bring the states of internal nodes during the test to the ATE system. Often, only those nodes that can be probed are desired, while those nodes that exist deeper in the hierarchy of the model (like internal IC nodes) may not be wanted. When extracting internal state data from the CAE system database, this differentiation must be observed.

- How is internal state data organized on the ATE system?
- What is the syntax of these data?
- What are the restrictions on characters?

B.3.6. Size Limitations

When using automated tools to produce tests on CAE systems, it is possible that much larger amounts of data may be produced than were previously (or humanly) feasible.

Here are some questions:

- For each type of file to be transfered, what are the limits on file size?
- For files that are to be compiled, what are the practical limits on the amount of data they contain?
- For files that are known to be large, is there a way to segment them into manageable pieces?
- If large files are to be transferred, does the transfer method have an acceptable error rate?

B.3.7. Summary

The fundamental message provided by this appendix, and indeed the whole book, is that successful test development requires planning and a strategy. The checklists provided here are without doubt incomplete. It is hoped that they will form a framework upon which each person, charged with the responsibility of developing tests by using CAE tools, can prevail.

Index

INDEX